# Saltmarsh Conservation, Management and Restoration

# Coastal Systems and Continental Margins

VOLUME 12

*Series Editor*

Bilal U. Haq

*Editorial Advisory Board*

M. Collins, *Dept. of Oceanography, University of Southampton, U.K.*
D. Eisma, *Emeritus Professor, Utrecht University and Netherlands Institute for Sea Research, Texel, The Netherlands*
K.E. Louden, *Dept. of Oceanography, Dalhousie University, Halifax, NS, Canada*
J.D. Milliman, *School of Marine Science, The College of William & Mary, Gloucester Point, VA, U.S.A.*
H.W. Posamentier, *Anadarko Canada Corporation, Calgary, AB, Canada*
A. Watts, *Dept. of Earth Sciences, University of Oxford, U.K.*

*The titles published in this series are listed at the end of this volume.*

# Saltmarsh Conservation, Management and Restoration

By
J. Patrick Doody
*National Coastal Consultants, Brampton, UK*

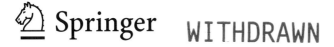 Springer    WITHDRAWN

J. Patrick Doody
National Coastal Consultants
Brampton
UK

*Series Editor:*
Bilal V. Haq
Bethesda
MD
USA

**Cover illustrations**
Left: *Salicoria europaea*. Top right: Dissected, grazed upper saltmarsh in Devon. Bottom right: Salt pan. Photographs taken by J. Patrick Doody.

ISBN 978-1-4020-4603-2      eISBN 978-1-4020-5748-9

Library of Congress Control Number : 2007933718

© 2008 Springer
No part of this work may be reproduced, stored in a retrieval system, or transmitted in any form or by any means, electronic, mechanical, photocopying, microfilming, recording or otherwise, without written permission from the Publisher, with the exception of any material supplied specifically for the purpose of being entered and executed on a computer system, for exclusive use by the purchaser of the work.

Printed on acid-free paper.

9 8 7 6 5 4 3 2 1

springer.com

*This book is dedicated to my wife who on some of our trips could be heard to say 'oh not another saltmarsh'. Despite this, we are still together after more than 30 years. Perhaps saltmarshes have a little extra after all!*

# Contents

**Foreword** . . . . . . . . . . . . . . . . . . . . . . . . . . . . . . . . . . . . . . . . . . . . . . . . . . . xiii

**Preface** . . . . . . . . . . . . . . . . . . . . . . . . . . . . . . . . . . . . . . . . . . . . . . . . . . . . . xvii

**Acknowledgements** . . . . . . . . . . . . . . . . . . . . . . . . . . . . . . . . . . . . . . . . . xix

1. **Introduction** . . . . . . . . . . . . . . . . . . . . . . . . . . . . . . . . . . . . . . . . . . . . . 1
   - 1.1 Development . . . . . . . . . . . . . . . . . . . . . . . . . . . . . . . . . . . . . . . . 1
   - 1.2 Sediments and Sedimentary Processes . . . . . . . . . . . . . . . . . . . . . . 1
   - 1.3 Saltmarsh Vegetation . . . . . . . . . . . . . . . . . . . . . . . . . . . . . . . . . . 3
     - 1.3.1 Tides and Transitions . . . . . . . . . . . . . . . . . . . . . . . . . . . 4
     - 1.3.2 Succession . . . . . . . . . . . . . . . . . . . . . . . . . . . . . . . . . . . 4
     - 1.3.3 The Nature of the Vegetation . . . . . . . . . . . . . . . . . . . . . 6
   - 1.4 An Ecosystem Approach . . . . . . . . . . . . . . . . . . . . . . . . . . . . . . . 7
     - 1.4.1 Productivity . . . . . . . . . . . . . . . . . . . . . . . . . . . . . . . . . . 7
     - 1.4.2 Internal and External Relationships . . . . . . . . . . . . . . . . . . . 9
   - 1.5 Geographical Location and Type . . . . . . . . . . . . . . . . . . . . . . . . . . 10

2. **Human Influences** . . . . . . . . . . . . . . . . . . . . . . . . . . . . . . . . . . . . . . . 17
   - 2.1 Introduction . . . . . . . . . . . . . . . . . . . . . . . . . . . . . . . . . . . . . . . . 17
   - 2.2 Traditional Management . . . . . . . . . . . . . . . . . . . . . . . . . . . . . . . 17
     - 2.1.1 Grazing . . . . . . . . . . . . . . . . . . . . . . . . . . . . . . . . . . . . . 17
     - 2.1.2 Reed Cutting . . . . . . . . . . . . . . . . . . . . . . . . . . . . . . . . . 18
     - 2.1.3 Samphire Gathering . . . . . . . . . . . . . . . . . . . . . . . . . . . . 19
     - 2.1.4 Saltmarsh 'Haying' . . . . . . . . . . . . . . . . . . . . . . . . . . . . 19
   - 2.2 Excavation . . . . . . . . . . . . . . . . . . . . . . . . . . . . . . . . . . . . . . . . . 19
     - 2.2.1 Turf Cutting . . . . . . . . . . . . . . . . . . . . . . . . . . . . . . . . . 20
     - 2.2.2 Sediment Extraction . . . . . . . . . . . . . . . . . . . . . . . . . . . . 20
   - 2.3 Summer Dykes, Grazing Marsh, Salinas and Rice Fields . . . . . . . . 20
     - 2.3.1 Summer Dykes . . . . . . . . . . . . . . . . . . . . . . . . . . . . . . . 21
     - 2.3.2 Grazing Marsh . . . . . . . . . . . . . . . . . . . . . . . . . . . . . . . 21
     - 2.3.3 Salinas . . . . . . . . . . . . . . . . . . . . . . . . . . . . . . . . . . . . . 22
     - 2.3.4 Rice Cultivation . . . . . . . . . . . . . . . . . . . . . . . . . . . . . . 23

|  |  |  |  |
|---|---|---|---|
| 2.4 | Enclosure and Habitat Loss – 'Land Claim' | | 24 |
| | 2.4.1 | Saltmarshes, Other Coastal Wetlands and Mosquitoes | 25 |
| | 2.4.2 | Enclosure and Drainage of Coastal Wetlands for Agriculture | 25 |
| | 2.4.3 | 'Warping' and Sediment Fields | 28 |
| | 2.4.4 | Enclosure for Infrastructure | 29 |
| | 2.4.5 | Summary of Enclosure | 30 |
| 2.5 | Other Influences | | 31 |

## 3. Nature Conservation. 32
- 3.1 Introduction . 32
- 3.2 The USA . 33
- 3.3 The Changing Scene in Europe . 34
- 3.4 Habitat Loss and Nature Conservation in the UK . 34
    - 3.4.1 Saltmarsh Enclosure in the Wash, South-East England. 36
    - 3.4.2 Erosion of Essex Saltmarsh, South-East England . 38
    - 3.4.3 Cardiff Bay, South Wales. 41
- 3.5 From Enclosure to Enlightenment and Realignment. 41
    - 3.5.1 'Protecting' Essex Saltmarsh. 42
    - 3.5.2 The UK 'Saltmarsh Squeeze'. 43
    - 3.5.3 Enlightenment . 43
- 3.6 The Rest of Europe . 45
    - 3.6.1 Europe – ICZM, Erosion and All That . 45
    - 3.6.2 Accommodating Change – 'Living with the Sea' . 46

## 4. States and Values. 48
- 4.1 Introduction . 48
    - 4.1.1 Driving Forces, Pressures, States, Impacts and Response (DPSIR) . 48
- 4.2 Physical States – Description. 49
    - 4.2.1 State 1 – Accreting. 50
    - 4.2.2 State 2 – Semi-Stable (Dynamic Equilibrium) . 52
    - 4.2.3 State 3 – Eroding . 55
- 4.3 Physical States – Values. 57
    - 4.3.1 Ecosystem Values. 57
    - 4.3.2 Economic Values . 58
    - 4.3.3 Cultural Values . 62
- 4.4 Vegetative States – Description . 64
    - 4.4.1 State 1 – Heavily Grazed . 65
    - 4.4.2 State 2 – Moderately Grazed. 66
    - 4.4.3 State 3 – Historically Ungrazed/Lightly Grazed . 66
    - 4.4.4 State 4 – Abandoned, Formerly Grazed . 67
    - 4.4.5 State 5 – Overgrazed . 69
- 4.5 Vegetation States – Values . 69
    - 4.5.1 Nature Conservation Values . 69

## 5. The Physical States ............................................. 74
- 5.1 Introduction ............................................... 74
- 5.2 Physical Trends ........................................... 75
  - 5.2.1 Processes Influencing the Physical State ................ 75
- 5.3 Values Associated with the Physical State
  of the Saltmarsh........................................... 75
  - 5.3.1 State 1 – Accreting Saltmarsh ....................... 77
  - 5.3.2 State 2 – Dynamically Stable........................ 78
  - 5.3.3 State 3 – Eroding Saltmarsh......................... 79
- 5.4 Summary – A Physical Model for Change .................... 80
  - 5.4.1 Rates of Accretion and Loss ......................... 81
- 5.5 Monitoring is an Essential Tool ............................. 83
- 5.6 Assessing the Need for Intervention ......................... 85
  - 5.6.1 Accreting – State 1................................. 85
  - 5.6.2 Dynamic Equilibrium – State 2 ...................... 86
  - 5.6.3 Eroding – State 3 .................................. 86
- 5.7 Approaches to Restoration................................. 87
  - 5.7.1 Moving Seaward, Creating New Saltmarsh.............. 87
  - 5.7.2 Protecting and Restoring Saltmarsh................... 88
  - 5.7.3 Moving Landward, Re-integration and Habitat Creation ... 89

## 6. Physical States, Restoration Methods........................... 90
- 6.1 Introduction .............................................. 90
- 6.2 Restoring Eroding (State 3) Saltmarsh, Moving
  Seaward or 'Staying Put'................................... 90
  - 6.2.1 Warping, Poldering ................................ 91
  - 6.2.2 Bay Bottom Terracing in the USA.................... 92
  - 6.2.3 Use of Dredged Material ............................ 93
  - 6.2.4 Reseeding and Other Vegetation Restoration ........... 94
  - 6.2.5 Planting Cord Grass *Spartina* spp. in the USA ........... 94
  - 6.2.6 Offshore Breakwaters .............................. 95
  - 6.2.7 Rip-Rap, Protecting the Eroding Edge................. 97
  - 6.2.8 Setbacks to Planting Native Spartina.................. 97
- 6.3 Restoring Saltmarsh – Moving Landward:
  'Re-integration with the Sea'................................ 98
  - 6.3.1 Restoring Tidally Restricted Saltmarshes in the USA...... 98
  - 6.3.2 The German Baltic Coast............................ 99
- 6.4 Re-creating Saltmarsh from Agricultural Land in England,
  'Abandonment' or 'Managed Realignment' ................... 99
  - 6.4.1 Abandonment...................................... 100
  - 6.4.2 Realignment....................................... 102
- 6.5 Conclusions .............................................. 106

## 7. Vegetation States ... 108
### 7.1 Introduction ... 108
### 7.2 Mechanisms for Change – Native Species ... 108
#### 7.2.1 Native Waterfowl in Europe ... 109
#### 7.2.2 Mammals ... 110
#### 7.2.3 Introduced Nutria in North America ... 110
#### 7.2.4 Snails and Other Invertebrates ... 111
### 7.3 Mechanisms for Change – Grazing by Domestic Livestock ... 112
#### 7.3.1 Changes in Plant Communities ... 112
#### 7.3.2 Changes in Rare Species ... 114
#### 7.3.3 Grazing and Breeding Birds ... 116
#### 7.3.4 Grazing by Domestic Stock, Effects on Avian Herbivores ... 117
#### 7.3.5 Grazing and Invertebrates ... 118
#### 7.3.6 Grazing and Sea Bass ... 119
### 7.4 Lesser Snow Geese (*Chen caerulescens caerulescens*) – A Conservation Dilemma? ... 120
#### 7.4.1 A Population Explosion ... 120
#### 7.4.2 Grazing Impacts on Saltmarsh ... 120
#### 7.4.3 Controlling Goose Numbers ... 122
### 7.5 Changing Biological Values ... 122
### 7.6 'Grazing' State Evaluation Model ... 124
#### 7.6.1 Effecting Vegetative Change – A State Evaluation Model ... 125

## 8. Grazing Management ... 126
### 8.1 Introduction ... 126
### 8.2 Assessing the Need for Change ... 127
#### 8.2.1 Historical Understanding ... 127
#### 8.2.2 Protecting Nature Conservation Values ... 128
#### 8.2.3 Assessing the Implications of Grazing ... 129
#### 8.2.4 Grazing Management in North America ... 130
### 8.3 Managing Grazing Levels ... 131
#### 8.3.1 Maintaining Ungrazed/Lightly Grazed 'Natural' Saltmarsh (State 3) ... 131
#### 8.3.2 Maintaining Moderately Grazed Saltmarsh (State 2) ... 132
#### 8.3.3 Maintaining Heavily Grazed Saltmarsh (State 1) ... 133
### 8.4 Modifying Saltmarsh Vegetation ... 133
#### 8.4.1 Historically Ungrazed or Lightly Grazed (State 3) and Moderately Grazed (State 2) Saltmarsh ... 133
#### 8.4.2 Reducing Grazing Pressure – Heavily Grazed (State 1) Saltmarsh ... 134
#### 8.4.3 Restoring Grazing on Abandoned (State 4) Saltmarshes ... 135
#### 8.4.4 Mowing as a Management Tool on Abandoned Saltmarshes ... 137
#### 8.4.5 Restoring 'Overgrazed' Saltmarsh ... 138
### 8.5 Conclusions ... 138

## Contents

**9. *Spartina*** .................................................................. 139
    9.1    Introduction ......................................................... 139
        9.1.1    The Nature of Colonisation ......................... 139
        9.1.2    Hybridisation ................................................ 141
        9.1.3    Pattern of Invasion ...................................... 142
        9.1.4    Rates of Sedimentation................................ 143
    9.2    World Domination ............................................. 144
        9.2.1    World Resources........................................... 145
        9.2.2    Spread in China, Australia and New Zealand .......... 145
        9.2.3    USA, Washington State and San Francisco Bay ........ 146
    9.3    Changing Perceptions........................................ 147
        9.3.1    Impacts on Bird Populations in the UK ............... 148
        9.3.2    Impacts on Amenity Beaches, North-West England ..... 148
        9.3.3    Problems in the USA .................................. 149
        9.3.4    Studies Elsewhere......................................... 150
    9.4    Methods of Control ........................................... 150
        9.4.1    Herbicides..................................................... 151
        9.4.2    Physical/Mechanical Control ...................... 152
        9.4.3    Grazing......................................................... 152
        9.4.4    Biological Control ....................................... 152
        9.4.5    Summary of Control Measures.................... 153
    9.5    *Spartina* spp. Friend or Foe? ........................... 154
        9.5.1    Control – Concerns and Costs ..................... 154
        9.5.2    'Natural Die Back' ....................................... 155
        9.5.3    Changing Patterns of Invasion – Great Britain .......... 157
        9.5.4    *Spartina* in North-West England, a Case of Succession .. 158
    9.6    Conclusion ......................................................... 161
        9.6.1    *Spartina anglica* – A Natural Component
                of Saltmarshes in the UK and Ireland?................. 161
        9.6.2    Friend or Foe ............................................... 162

**10. Conclusions** ............................................................. 165
    10.1    Introduction ...................................................... 165
    10.2    Time and Tide Wait for No One........................ 165
        10.2.1    Can Saltmarshes Keep Pace with Sea-Level Rise? ..... 167
    10.3    Saltmarshes and Saltmarsh Restoration............. 168
    10.4    Southern North Sea ........................................... 168
        10.4.1    Will it All Come Out in the Wash?................ 169
        10.4.2    Realignment in Belgium............................. 171
        10.4.3    The Wadden Sea.......................................... 172
    10.5    Restoration in North America ........................... 173
        10.5.1    The State of Louisiana and the Mississippi Delta ...... 174
        10.5.2    San Francisco Bay ..................................... 175
        10.5.3    Restoration in Canada ................................ 175

| | 10.6 | The Wider Role – Management and Restoration............. 175 |
| | | 10.6.1 Approaches to Restoration in the Wadden Sea, the Netherlands and Germany..................... 176 |
| | | 10.6.2 Depoldering, the Delta Region of the Netherlands..... 177 |
| | | 10.6.3 The Situation in the UK, Winning Hearts and Minds ... 177 |
| | | 10.6.4 The Mediterranean, Sediments and Deltas........... 179 |
| | | 10.6.5 The Ebro Delta............................... 180 |
| | | 10.6.6 The Venice Lagoon............................ 181 |
| | 10.7 | The Future.............................................. 181 |
| | | 10.7.1 A European Initiative........................... 182 |
| | | 10.7.2 A Historical Perspective......................... 183 |
| | | 10.7.3 A New Perspective on Saltmarsh Conservation?....... 184 |
| | 10.8 | What Does the Future Hold?............................. 186 |

**Appendix – English and Latin Names**............................. 187

**Appendix – A Few Useful Web Sites**............................. 190

**References**..................................................... 193

**Index**.......................................................... 213

**CD-ROM included inside back cover**

# Foreword

The saltmarsh is one of a suite of coastal habitats that depend on the interaction of the land with the sea. It is essentially a plant community, which develops on the upper shore in sheltered tidal inlets and bays. This book deals mainly with the saltmarshes of temperate parts of the northern hemisphere. Whilst these are amongst the most natural of habitats exhibiting primary succession, it emphasises the impact that human actions have had over the centuries. The effects range from habitat loss due to human intervention to modification of the structure of the vegetation by grazing of domestic stock. As these changes take place saltmarsh biodiversity, ecosystem values and use to society also change.

Enclosure for infrastructure or agricultural use completely destroys the habitat and results in the total loss in values associated with it. Grazing on the other hand changes the values rather than destroying them. Inevitably, these modifications involve a 'trade-off' between one environmental value and another and with socio-economic values. In some areas, the changes have reached the point where the saltmarsh no longer fulfils its 'natural' role in the functioning of the ecosystem in which it exists. This has both environmental and socio-economic consequences, in some areas adversely affecting local human populations. In these circumstances, changes in management and/or restoration may be essential for the long-term sustainability of the ecosystem as a whole and its ability to provide 'goods and services' to the human population. The book looks at the conservation, management and restoration of the habitat and the 'trade-offs' associated with changes in policy, in ten chapters, as follows:

- Chapter 1 describes the nature of the saltmarsh habitat and its role in the wider coastal ecosystems
- Chapter 2 illustrates the way in which human actions have modified the habitat
- Chapter 3 reviews the changing human perception of saltmarshes
- Chapter 4 gives detailed information on the range and variation of the habitat and the values attached to it
- Chapter 5 relates Chapters 2, 3 and 4 to each other. It assesses the extent to which changes in management affect the socio-economic and environmental values of the *physical* state of the saltmarsh

- Chapter 6 relates Chapters 2, 3 and 4 to each other. It assesses the extent to which changes in management affect the environmental values of the *vegetative* state of the saltmarsh
- Chapter 7 looks at the methods of achieving changes in the physical state of the saltmarsh in relation to the balance between accretion and erosion
- Chapter 8 looks at the methods of achieving changes in the vegetative state of the saltmarsh, mainly in relation to alteration in grazing pressures
- Chapter 9 reviews the role of *Spartina* spp. as a natural and introduced component of saltmarshes
- Chapter 10 provides a more general review at the role of saltmarshes, in coastal systems, especially in the face of global climate change

This book only deals with temperate saltmarshes in the northern hemisphere. Similar habitats exist in the southern hemisphere and there is an important literature for Australia and New Zealand, and for China and Japan. Where directly relevant, these sources are included. Mangroves are not covered. Consideration is given, where appropriate, to some of the secondary habitats (those derived from saltmarshes, such as rice fields, salinas and coastal grazing marshes). For an introduction to these habitats, consult Doody (2001, Chapter 11).

It provides an introductory framework to the problems of saltmarsh conservation, management and restoration. It gives the reader background information on the issues and pointers to their possible solutions. Descriptions of the trends and trade-offs help to identify different policy options.

The author has searched the Internet, and web site URLs used as sources of information are included in the text. Many individual countries support web sites specifically dealing with coastal ecosystems, including saltmarshes. Australia, for example, has general coastal information provided by the Australian Government, Department of Environment and Heritage. See the 'Coasts and Oceans' web site home page (http://www.deh.gov.au/coasts/index.html) which includes an introduction to Australia's saltmarshes provided by Adam (1995).

Foreword    xv

The book also includes an extensive list of references.

Ravenglass Estuary, Cumbria, northern England. Fringing saltmarshes and mudflats

# Preface

Coastal habitats provide the link between the land and the sea. They are dynamic, combining to form ecosystems of great complexity and significant areas for wildlife. Visitors, painters and musicians treasure these landscapes. They also provide locations for significant economic activity. This may include fisheries, providing food and shelter for some species of commercially exploited fish stocks. The habitats themselves are a buffer to tides and wave action, which may be particularly important in areas where relative sea level is rising, and during storm periods. Managing these assets in the face of continuing pressure from human populations on a sustainable basis is a major task.

This book follows up and expands upon the more general book *Coastal Conservation and Management: An Ecological Perspective* (Doody 2001).* It is the first in a series looking at the main coastal habitats – saltmarshes, sand dunes and sandy shores, coastal vegetated shingle and shingle shores and sea cliffs. A description of the natural development of each habitat provides the basis for consideration of the influence of human actions. The different states in which the habitats exist are reviewed against the pressures exerted upon them. Options for management are considered and the likely consequences of taking a particular course of action highlighted. These options include the traditional approaches to management (for the conservation of wildlife and landscapes) as well as habitat restoration. The way the value of the areas change under different management regimes is discussed from both a socio-economic and environmental perspective.

Some of the information and ideas included in this book are developed from the Internet Guide 'Coastal Habitat Restoration, Towards Good Practice'. Prepared as part of the European Life Nature Project 'Living with the Sea' (http://www.english-nature.org.uk/livingwiththesea/project_details/default.asp) by the author.

---

* Doody, J.P., 2001. *Coastal Conservation and Management: an Ecological Perspective.* Kluwer, Academic Publishers, Boston, USA, 306 pp. Conservation Biology Series, No. 13.

Some of the sites mentioned in the text are shown in the following figures:

- European sites (excluding Great Britain) Figure 5, p. 12
- Sites in Great Britain, Figure 6, p. 13
- Sites on the east coast of England, Figure 49, p. 101
- Sites in Essex, Figure 73, p. 182
- Sites in the USA, Figure 61, p. 147

# Acknowledgements

This book represents a synthesis of research and information derived from the work of a large number of scientists, managers and policy advisers over the last 70 years or so. The early pioneering ecological studies of people such as Carey & Oliver (1918), V.J. Chapman, in the 1930s and 1940s and the work of Derek Ranwell in the 1960s and 1970s set the foundation for our present day understanding of the saltmarsh habitat. Recently, others such as Adam (1990), Allen & Pye (1992), and Packham & Willis (1997) have set the habitat in a wider ecological setting. Carter (1988), Carter & Woodroffe (1994), Pethick (1984) and Bird (1984) have provided the geomorphological context. I had the privilege of meeting and working with some of them.

Former colleagues in the Nature Conservancy Council frequently enhanced my visits to the UK coast. In Europe, officers and members of the Coastal Union (European Union for Coastal Conservation) led expeditions to some of the remoter parts where saltmarshes and other coastal wetlands retained many of the natural features. I am grateful to Albert Salman (Secretary General of the Coastal Union) for his friendship and companionship, especially on our trips to eastern Turkey and Albania in the 1990s.

My thanks to Fiona Burd and others, who over many long hours in the field, carried out the survey of saltmarshes in Great Britain (Burd 1989). I am pleased that policy makers and developers no longer see saltmarshes and other tidal land as 'wastelands'. In England, at least, Mark Dixon and John Pethick helped give impetus to saltmarsh re-creation, through managed realignment. This, in turn, gave support to nature conservation arguments for the protection and management of the surviving saltmarshes in the UK as a whole.

I hope that this little book will help others recognise their immense value, both in nature conservation terms and for the values that accrue to humans.

Thanks to Kees Dijkema, with whom I worked on the original inventory of European saltmarshes (Dijkema 1984), and who kindly agreed to review the book for me. Finally, thanks to Professor John Allen for his rigorous review and helpful comments on the style and content of the book.

Some saltmarsh specialists, from left to right, Alan Gray, Derek Ranwell (deceased) and Kevin Charman, in a 'field' of *Spartina anglica* on the Ribble Estuary, north-west England, 1983.

# Chapter 1
# Introduction

## Saltmarshes and Ecosystems

### 1.1 Development

Lying at the margin of muddy/sandy foreshores and the land, saltmarsh has characteristics of both the marine and terrestrial environment. Many ecological textbooks describe the 'natural' saltmarsh succession. Some of the early classic ecological studies were of the saltmarsh vegetation on the North Norfolk coast (see, e.g. Chapman 1938, 1939, 1941) or the Dovey Estuary in west Wales (Yapp et al. 1917). These early studies helped provide the basis, not only for describing saltmarsh development, but also as part of a wider understanding of vegetation succession. The relationship between tidal inundation and the parallel spatial zonation, often observed within the habitat, was established. The importance of soil properties, plant species strategies, community structure and function as well as trophic relationships and energy flows were also identified and further elaborated (e.g. Ranwell 1972; Long & Mason 1983). Other work relates the vegetation to its geomorphological origins (see in particular Adam 1990).

These studies helped unravel the way that saltmarshes develop. This book stresses the fact that saltmarshes usually occur in a complex series of relationships, within a wider coastal ecosystem. The action of ice, water, wind and biotic activity help to create this and other habitats. Saltmarshes, sand dunes or shingle structures come together to form mosaics of interacting ecosystems in estuaries, coastal deltas and embayments. These relationships change over time at scales stretching from a few minutes (storms and waves), to hours (tidal movement) and weeks or years (sediment supply, sea-level change). Geomorphological textbooks describe these processes in some detail, e.g. Bird (1984, 1985); Carter (1988); Carter & Woodroffe (1994); Pethick (1984).

### 1.2 Sediments and Sedimentary Processes

Underlying the development of saltmarsh is the movement and deposition of sediment. Three main sediment sources provide the material, namely:

1. Erosion of elevated land, sediment transported by rivers to the sea;
2. Erosion of sea cliffs, sediment transported by tides, waves and storms and moved along the shore by longshore drift (longshore drift is the movement of material along a coast by waves, which approach at an angle to the shore);
3. Reworking of subtidal offshore banks, sediment moved by coastal waters.

Reworking of existing coastal habitats within estuaries, deltas and embayments, also releases sediment for transport (Figure 1).

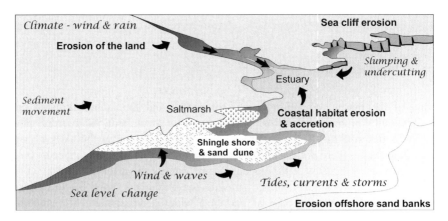

**Figure 1** Processes affecting the supply and/or movement of sediments in the coastal zone

The combination of sediment supply and physical location help determine the location and scale of saltmarsh development. The most extensive saltmarshes normally exist within the confines of estuaries, deltas and other embayments, where the deposition of sediment to form tidal flats takes place. The processes involved are complex and begin with the settlement of sediments to form sand and mud flats. Wave-dominated shores typically have sand flats; tide-dominated ones mud flats. Enclosing or protective spits, barriers or foreshore slow tides and wave action. Suspended sediment in the water column settles out at or near high water, at the point when tidal velocities are at their slowest. When the combination of tidal movement and wave action are insufficient to erode the settled material, accretion takes place (Allen 2000; Postma 1961). As the sediment accumulates, the rising shoreline becomes free from tidal inundation for increasingly longer periods. This allows the establishment of salt-tolerant plant species, which ultimately form saltmarsh.

Figure 2 is a simplified picture of the physical variation in saltmarsh type. A more expansive classification of the saltmarshes of estuaries, for example, also includes open embayments, estuarine fringing and estuarine back-barrier saltmarshes (Pye & French 1993). The figure excludes 'perched' saltmarshes (on sea cliffs) and 'beach-head' saltmarshes (on exposed rocky shores), where the communities are dominated by typical salt-tolerant species but receive little or no sediment (Doody 2001).

1.3 Saltmarsh Vegetation

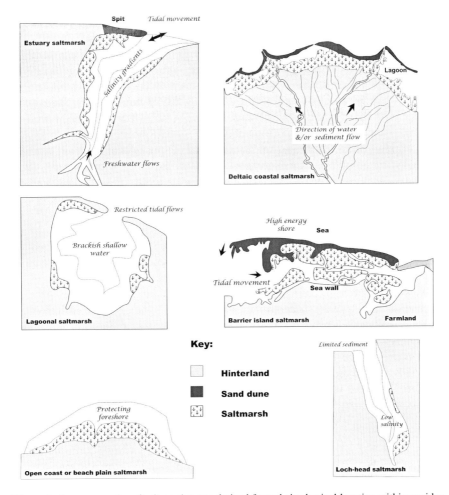

**Figure 2** Some examples of saltmarsh types derived from their physical location within a wider geomorphological system

## 1.3 Saltmarsh Vegetation

The supply of sediment and the ability of saltmarsh plants to withstand tidal inundation are key factors in determining whether saltmarshes develop or not. Saltmarshes become established where sediment builds up to a point where pioneer plants can become established. This early saltmarsh vegetation is composed of a small number of salt-tolerant (halophytic) species and is most often found on intertidal surfaces where the roots of germinating plants are left undisturbed by tidal action for several days (Ranwell 1972). The degree of shelter from wave action is also important. Fine sediments ($>20\%$) in the surface layers of the flats improves the shear strength and with it the settlement and survival, for example, of pioneer Long-spiked Glasswort (*Salicornia dolichostachya*) (Houwing et al. 1999). Under favourable conditions, extensive tidal flats may have large areas of saltmarsh along their margins.

### 1.3.1 Tides and Transitions

Tidal regimes differ, depending on the nature of the position of the coast from an amphidromic point\* and the alignment of the sun and moon with the earth. High (spring) tides occur when the earth, sun and moon are more or less in alignment and the gravitational pull is at its greatest (new and full moons). Low (neap) tides occur when the sun and the moon are at right angles to the Earth, resulting in minimal gravitational attraction (first and last quarter moons). The gradual switch between spring and neap tides occurs approximately every two weeks, over each monthly tidal cycle. Higher spring tides occur during the equinoxes, when the Sun is overhead at the equator (around 20 March and 23 September).

Most tides are semi-diurnal, i.e. there are two high and two low tides per day. The period between each successive tide is 12 hours 25 minutes so that each day high tide is 50 minutes later than the previous one. This is important for anyone studying saltmarshes, as the incoming (flood tide) can be very dangerous. Tide tables provide a prediction of the time and height of daily tides for many parts of the world, especially around major ports. There are other factors, which determine the state of the tide. The ocean waters move in response to the gravitational pull of the moon and the sun and the rotation of the earth. Relatively small ocean waves, created by these forces, become amplified as they approach a continental shelf and flow into coastal embayments. The length and depth of the tidal basin help to determine the final height of the wave and hence the tidal range at a particular location. For more information on the various factors influencing the tides around the world see Parker (2005).

### 1.3.2 Succession

Vegetation succession takes place as the surface of the saltmarsh rises in response to sediment deposition and stabilisation. In temperate waters, this is usually between mean high water of spring tides (MHWS) and mean high water of neap

---

\*The amphidromic point is the point in an ocean basin around which waves rotate. Tidal movement at the point is zero, but increases as the water movement radiates outwards towards the coast. The result is different tidal regimes in different parts of the world. The highest tidal ranges are found in the Bay of Fundy, Nova Scotia, Canada and in the Bristol Channel, south-west Britain. The average for the former is 12 m although exceptionally they may reach 16 m. In the latter, the range is in excess of 14.5 m, and the 'Severn Bore', which is generated by the tide as it moves upstream, can travel at 16 km per hour.

## 1.3 Saltmarsh Vegetation

tides (MHWN). As the height of the surface increases, progressively fewer tides cover the vegetation. The rate of deposition declines with increasing distance from the main channels and creek banks. This in turn results in higher saltmarsh communities receiving less sediment and a slower rate of accretion. Measurements in the Severn Estuary, UK show this to be:

1. 12.1 mm per year on the lowest saltmarsh surface, covered by most high tides;
2. 06.4 mm per year on the mid-level saltmarsh surface;
3. 02.3 mm per year on the upper saltmarsh surface, only covered by the tide on high spring tides (French 1997).

Although the sequence of succession cannot be derived from the vegetation mosaic (Adam 1990), the species composition changes with elevation. Tolerance to submergence in salt water is a prerequisite for colonisation at lower levels with relatively few species occupying distinctive zones. At higher elevations and well-drained locations (levees), less salt-tolerant plants occur and include a greater range of species (Figure 3).

Competition seems to be more important in determining the plant communities at these higher levels. A diverse community of plants and animals develops towards the upper levels of the saltmarsh, which can include species-rich transitions to fen and other freshwater communities. The factors determining zonation and succession across a saltmarsh are discussed in some detail in Adam (1990, pp. 49–57), Gray (1992) and Packham & Willis (1997, pp. 107–114).

**Figure 3** An example of a simple succession on a saltmarsh as the height of the marsh increases away from the tidal creek (Dee Estuary, Merseyside, UK). Common Saltmarsh-grass (*Puccinellia maritima*), Sea-purslane (*Atriplex portulacoides*) and Red Fescue (*Festuca rubra*) at progressively higher levels from the creek bottom outwards. The site is largely ungrazed and in addition to the sward of tall grasses including *Festuca rubra*, Sea Couch (*Elytrigia atherica*) a species less tolerant of daily submergence by seawater, stretches into the middle distance

### 1.3.3 The Nature of the Vegetation

As saltmarsh develops patterns of variation occur, which reflect the tidal regime, sediment availability and sediment type. Prevailing climatic conditions add another dimension helping to create patterns discernable at three scales:

1. Microscale – vegetation patchiness on an individual marsh based on the presence of individual species (centimetres to metres for shrubby plants);
2. Mesoscale – bands of vegetation based on transitions across the marsh in response to tidal inundation. Mosaics superimposed on these transitions result from the development of creeks around saltpans or because of changes in the thickness of the clay layer, grazing patterns and/or rotting tidal litter. Within estuary, gradients determined by mixing tidal waters and freshwater river flows. Changes in the rate of sea-level rise alter the height in the tidal frame at which the saltmarsh reaches equilibrium. Scales range from tens to hundreds of metres;
3. Macroscale – involving regional variation such as exposure to westerly climates, continental patterns correlated with latitude or a combination of the two. Variable scales, i.e. from a few to several hundred kilometres.

These different scales combine to help create a pattern of vegetation distribution, which reflects the physical location, sediment availability and prevailing climatic conditions. Adam (1990) provides a modified macroscale worldwide classification, including the following geographical groups:

1. Arctic
2. Boreal
3. Temperate:

    (a) European
    (b) Western North American
    (c) Japanese
    (d) Australasian
    (e) South African

4. Western Atlantic Group
5. Dry coast
6. Tropical

There are many regional variations and Adam (1990, p. 155 ff) describes these in detail. This book is concerned mostly with the Boreal, Temperate and Western Atlantic saltmarshes of the northern hemisphere.

Climatic factors, such as the wind and rain, may act in the short term to moderate the type of vegetation developing in a particular location. Freshwater flows onto the marsh help to create biologically diverse transitions between saline and freshwater habitats. Biotic features including natural herbivores such as snails (e.g. *Hydrobia ulvae*) and geese, e.g. the Lesser Snow Goose (*Chen caerulescens caerulescens*), can alter the plant species composition. Taken together, these and other factors help to create a vegetation mosaic of some complexity.

## 1.4 An Ecosystem Approach

Saltmarshes are complex systems. The processes that combine to facilitate their development, act at a range of scales and over very different time intervals. Understanding these relationships provides the basis for identifying sound strategies for management.

Many of the factors determining whether a saltmarsh develops or not, operate on a scale beyond the limits of the saltmarsh itself. Sediment delivery and movement (Figure 1), help to define this wider ecosystem scale. Twice-daily cycles of the tide transport sediments across the shore. Tidal processes both determine and limit the areas within which, deposition takes place and saltmarshes become established. The extent and speed of sediment accumulation depends on a number of factors, including:

- Tidal range or wave energy;
- Sediment availability, type, method of deposition and cohesion;
- Degree of protection (from storms and wave action);
- Tidal flat topography and extent;
- Distance from coast/main estuary channel;
- Distance from creek banks.

The movement of saltmarsh landward or seaward also depends on whether the coast is rising (emerging) or falling (submerging) relative to sea level. These act over much longer timescales measured in decades or even centuries. Given sufficient sediment and an environment where deposition can take place, salt-tolerant plants establish themselves within the intertidal zone. Once established, a saltmarsh can grow vertically until it reaches a height above which the tides no longer cover it. The faster the rate of sea-level rise the lower the equilibrium position in the tidal frame. Horizontal limitations occur because of tidal scour, or other factors inhibiting lateral growth, which can also include sea-level rise.

The tidal range is significant in determining the direction of movement of sediments. The higher the range, the greater the tendency for the sediment to be driven landwards into 'coastal bays', to form 'estuarine', 'lagoonal', 'barrier island', 'open coast' or 'loch-head' saltmarshes (Figure 2). These types tend to be associated with macrotidal (>4 m tidal range) to mesotidal coasts (2–4 metres tidal range) that predominate around the margins of the major seas. Microtidal coasts (<2 m tidal range) tend to have wave- or river-dominated sedimentary regimes where extensive 'deltaic' saltmarshes develop (Figure 2), such as in the enclosed seas of the Mediterranean and Baltic Seas.

### 1.4.1 Productivity

Individual saltmarshes (especially those dominated by *Spartina* spp.) have high levels of primary productivity both above and below ground. The relationship between the two, based on estimates of the net total primary production of Smooth

Cord-grass (*Spartina alterniflora*) in a Georgia saltmarsh suggests that the belowground component is significant. The comparative figures in grams of carbon per $m^2$ per year are:

- Net aboveground production 1,421 for tall *S. alterniflora* and 749 for short;
- Net belowground production 872, tall *S. alterniflora* and 397 for short.

The short forms of *S. alterniflora* reached the productivity levels of the tall forms when nitrogen fertiliser was added (Dai & Wiegert 1996). Similar figures for net aboveground aerial production (Smalley technique) were:

- Tall *Spartina alterniflora*, 1,487 g per $m^2$ per year;
- Short *S. alterniflora*, 654 g per $m^2$ per year;
- Saltmeadow Cord-grass (*Spartina patens*) 1,147 g per $m^2$ per year;
- Saltgrass (*Distichlis spicata*) 785 g per $m^2$ per year (Roman & Daiber 1984).

Comparable figures for net above ground annual primary productivity using the same technique from saltmarshes in Europe were 450–500 g per $m^2$ per year (Boorman and Ashton 1997) similar to figures from France and the Netherlands. The lower figures may be the result of sampling communities with a variety of species exhibiting very different productivity rates in the sites studied in Europe. These have much lower overall values when compared with studies of single species such as *Spartina* spp. (Boorman 2003).

Overall, other belowground productivity values ranged from 80 g of carbon per $m^2$ per year for creek-head *Spartina alterniflora* to a high of 1,690 g of carbon per $m^2$ per year for Saltmeadow Rush (*Juncus gerardii*) in Maine. The mean for all plant stands was 650 g of carbon per $m^2$ per year. The average carbon content of the macroorganic matter (MOM) was 35.3%, corresponding to 1,850 g of dry weight per $m^2$ per year (Gallagher & Plumley 1979).

These studies also showed that there are three different types of belowground productivity profiles in Atlantic coastal marshes as determined by MOM:

1. Uniform with depth, notable example creek-bank *Spartina alterniflora* in the southern part of the coast;
2. High MOM concentration at the surface, decreasing with depth. This, the most common type of profile, shown by *Spartina patens*, *S. alterniflora* from the high marsh on the southern coast (Georgia), and creek-bank *S. alterniflora* from the northern part of its range (Maine);
3. A third type of profile had a large rhizome mat at 15–20 cm below the surface. Big Cord-grass (*Spartina cynosuroides*) and Common Reed (*Phragmites australis*) had a low biomass at the surface, a higher biomass just below the surface and a low concentration at depth (Gallagher & Plumley 1979).

The extent to which the exported material is utilised by other components of the ecosystem is complex. A commonly held view is that saltmarshes, especially *Spartina* saltmarshes, make a significant contribution to estuarine productivity. However, whatever the magnitude of primary production, the energy potentially available to other organisms in the wider ecosystem depends on the net primary production.

The studies referred to above attempt to determine the net primary production, which is the energy incorporated into the habitat through photosynthesis, less that used in respiration. For Common Cord-grass (*Spartina anglica*), at least, respiration may represent some 70% of the total biomass created through primary production (Long & Mason 1983). However, estimates of the net aboveground primary production in saltmarshes can vary widely according to the techniques of measurement (Linthurst & Reimold 1978). Thus, the high estimates of primary productivity attributed to saltmarshes, require moderation in relation to their potential contribution to the overall estuary or coastal embayment productivity.

Some saltmarshes may be net importers of material for some of the time, though annually they are still net exporters. A study of *Spartina anglica* in a Suffolk estuary showed, for example, that the export of 15–20% of the annual net primary production was as particulate matter; 70% to the estuary and the rest deposited on the strandline (Jackson et al. 1986).

Overall, it may not be possible to be definitive about the contribution saltmarshes make to the overall energy flow within an individual estuary or other coastal embayment (Adam 1990). However, although the value of saltmarshes as net exporters to the estuary ecosystem may not be as significant as suggested by some studies, they still represent an important component in the energy flow.

## *1.4.2 Internal and External Relationships*

Once established, there are many relationships both internal and external. The habitat also interacts with other habitats and species within the wider tidal environment. These interactions arise at many levels and involve some of the following:

- Individual plants support a range and diversity of invertebrate animal species;
- Saltmarsh plants provide food for herbivores such as ducks and geese;
- The vegetation provides shelter for species visiting tidal areas. including nursery areas for fish;
- They contribute to the circulation of nutrients acting as both a sink and a source;
- They have biologically diverse transitions to other habitats;
- They are dynamic, responding to changes in storms, wave action and sea-level change.

These interactions are many and complex. Different plants provide a range of food and shelter opportunities for animals such as insects, crustaceans and breeding birds. In the pioneer zone, a small number of animal species occur at high densities. Further up the saltmarsh, but below mean high water, the majority of the arthropod species are crustaceans; above this zone, the majority are insects. Birds, such as seed-eating Redpoll (*Carduelis flammea*) feed along the strandline on seeds derived from the saltmarsh plants. The fauna of the mid-marsh is generally much richer and largely made up of terrestrial species. Many of these are phytophagous (plant-eating) insects, which are dependent on specific

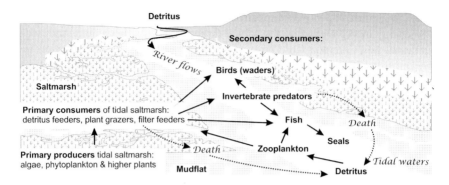

**Figure 4** Highly simplified diagram, showing the main trophic pathways in an estuary, in relation to saltmarsh

host plants or even parts of the host plants. The flowers of saltmarsh plants provide nectar for bumblebees and adult Lepidoptera. Some of these interactions are considered later in relation to the 'values' attached to saltmarsh vegetation as a habitat for animals (Chapter 3).

A range of primary producers provides the foundation for a food web within the saltmarsh. Herbivores, such as migrating ducks and geese, feed directly on the leaves of grasses and other plants. Plant detritus, derived from the dead and decaying saltmarsh plants, material produced in the tidal waters, together with that brought in by river flows and tidal inundation, provide food for a range of small, but abundant invertebrate animals. These become prey for wading birds, which are often the most visible manifestation of the high productivity of an estuarine ecosystem. Fish also feed on the plants and animals in the water column and are, themselves, preyed upon by other animals such as seals. These relationships operate both within the saltmarsh, and between it and the surrounding ecosystem (Figure 4).

A more detailed food web from a study of a salt meadow in New Zealand reproduced in Ranwell (1972, p. 123) and Packham and Willis (1997, p. 143), provides an illustration of the complexity of these interactions. Understanding this complexity is an important element in deciding on the best approach to the conservation, management and restoration of the habitat.

## 1.5 Geographical Location and Type

Saltmarshes occur throughout the world where conditions are suitable. There is no worldwide inventory of saltmarshes, though there are several papers, which provide summary data (Table 1).

Some of the larger saltmarshes in Europe occur in the tidal areas in and around the southern North Sea where meso- to macrotides occur. Extensive areas lie within

## 1.5 Geographical Location and Type

**Table 1** Area of saltmarsh for selected regions of the world

| Region | Tidal marsh area (ha) | Reference |
|---|---|---|
| Gulf Coast (North America) | 988,000* | Field et al. (1991) |
| Atlantic coast (North America) | 500,000–600,000* | Field et al. (1991) |
| Pacific coast (North America) | 44,000 | Field et al. (1991) |
| Great Britain | 45,337 | Burd (1989) |
| Western Europe | 95,000 | Dijkema (1990) |
| Wadden Sea | 39,680 | Bakker et al. (2005) |
| Southwest Atlantic coast (Brazil, Uruguay and Argentina) | 213,300 | Isacch et al. (2006) |
| China (*Spartina anglica*) | 39,000** | Chung C-H (1990) |

* Probably includes brackish to freshwater marshes; **this figure greatly underestimates the current area of saltmarsh, since Smooth Cord-grass (*Spartina alterniflora*) also rapidly invaded many coastal areas in China between 1993 and 2001 (From Zhang et al. 2004)

the Wadden Sea, though much of the mainland coast is composed of artificial saltmarshes, promoted by creating 'sediment fields' (Chapter 2). An example of a 'natural' saltmarsh on the barrier of Skallingen peninsula, Denmark has an area of 3,100 ha. Other sites include the Wash, England, which with 4,162 ha accounts for nearly 10% of the habitat in Great Britain (Burd 1989).

Perhaps the largest estuarine saltmarshes in Europe are those of the Odiel Marshes, Andalucía, Spain with an estimated area of 7,158 ha (Castillo et al. 2002). The size of this area is partly attributable to invasion by an introduced alien species into Europe. Dense-flowered Cord-grass (*Spartina densiflora*), which is dominant (>75% cover) in 18% of the saltmarsh (Mateos Naranjo et al. 2006), is thought to have been introduced from South America in the sixteenth century (Castillo et al. 2000). Saltmarshes of some size also occur in association with deltas in the microtidal seas, notably in the Mediterranean. In the Venice Lagoon, for example, there are some 3,700 ha of saltmarsh (Silvestri et al. 2003).

It is worth remembering, when considering the size of the saltmarshes along the European coast, that they were much more extensive than today. Relative to those in America they are an order of magnitude smaller, with large areas enclosed for 'land-claim' from the Roman period onwards. For example, in the Severn Estuary, UK alone the area of saltmarsh and transitions to brackish marsh amounted to around 100,000 ha, prior to enclosure (Allen pers. comm.).

The location of the smaller saltmarshes is closely allied to rocky coasts where embayments are restricted, limiting the tidal area suitable for saltmarsh establishment. Deeply incised, 'drowned' river valleys (rias) in the southwest and sea lochs in Scotland, Ireland and Scandinavia may also have narrow saltmarshes along their margins.

The distribution in Europe (Figure 5) and Great Britain (Figure 6) shown on the following pages is illustrative of the general trend for the larger saltmarshes to be associated with estuaries in the southern North Sea and the Irish Sea and deltas in the Mediterranean, respectively. Size is important for the role that the habitat plays in the overall functioning of the ecosystem, dealt with in later chapters.

**Figure 5** Saltmarsh distribution in Europe derived from Dijkema (1984), with additional information. The figure also shows the approximate location of the European mainland sites referred to in the text

The range of vegetation types in Europe spans the boreal, temperate and the dry coast types of Adam (1990). Saltmarshes in the north tend to be small, with short, often sheep-grazed swards including Glasswort (*Salicornia europaea*), Sea Thrift (*Armeria maritima*), Sea Arrowgrass (*Triglochin maritima*) and Saltmarsh Flat-sedge (*Blymus rufus*), as for example in northern Scotland (Figure 7). This community often includes free-living fucoids and is characteristic of the northern geographical zone, within the distribution of British saltmarshes (Adam 1978, 1981).

Temperate saltmarshes may be grazed or not and in the absence of enclosure show a succession from pioneer to transitional brackish and freshwater communities. In the UK through to southern Europe, there is a gradual change from frost-tolerant to warm-loving species. Many species widely distributed in the south reach their northern limit in the UK. For example, *Atriplex portulacoides*

## 1.5 Geographical Location and Type

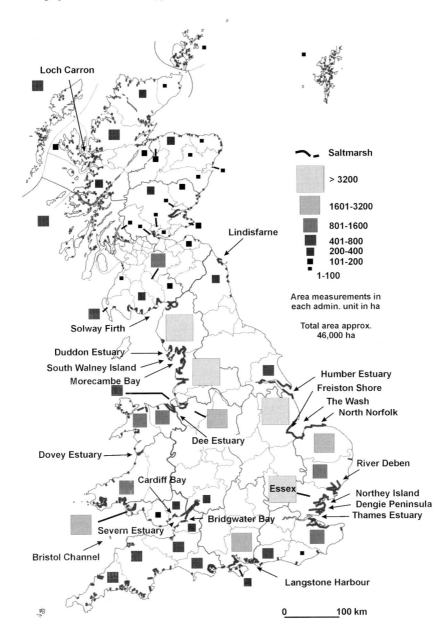

**Figure 6** Detailed survey of the saltmarshes in Great Britain (from Burd 1989). The figure shows the approximate location of the main sites referred to in the text

**Figure 7** A 'Loch-head' saltmarsh, typical of northern latitudes, in late summer, with a conspicuous covering of flowering Sea Thrift (*Armeria maritima*). Loch Carron in Scotland

is often dominant in ungrazed saltmarshes in England and Wales, but absent from most of Scotland. Small Cord-grass (*Spartina maritima*) is a native and often conspicuous pioneer species in many southern estuaries, especially in the centre of its range in southern Spain. Further north plants are smaller and restricted to southern England and the southern Netherlands, where it is sparsely distributed.

Mixed communities with *Salicornia spp.* and *Atriplex portulacoides* occur in and around estuaries in Portugal and south-west Spain (Figure 8). *Spartina maritima* is often the main pioneer species, though various species of Glasswort (*Salicornia*) are also present.

In the Mediterranean hypersaline conditions often occur. Saltmarshes therefore include various species of *Salicornia*, especially the shrubby Perennial Glasswort (*Sarcocornia perennis*) and the highly salt and drought tolerant, Glaucous Glasswort (*Arthrocnemum macrostachyum*). The latter species can be extensive on shallow upper tidal flats (Figure 9). The review (Dijkema 1984), carried out under the auspices of the Council of Europe in 1983/84, provides an account of the distribution of saltmarshes in Europe. It includes details of the plant cover, invertebrate interest and conservation of the habitat. It describes the geography of saltmarshes and their species in Section 2.3 with more detailed vegetation descriptions for the Arctic, North Atlantic, Central Atlantic, South Atlantic and West Mediterranean.

In North America, on the shores of the western Atlantic from southern Canada (Quebec and Newfoundland) to the Gulf of Mexico (Florida and Texas), saltmarshes are present along a substantial portion of the coast. These are characterised and often dominated by *Spartina alterniflora* (Chapter 9), the West

1.5 Geographical Location and Type

**Figure 8** A view across the Sado Estuary, Portugal. The saltmarsh community includes Purple Glasswort (*Salicornia ramosissima*), Annual Sea-blite (*Suaeda maritima*), *Atriplex portulacoides*, Glaucous Glasswort (*Arthrocnemum macrostachyum*), *Puccinellia maritima* and patches of Common Sea-lavender (*Limonium vulgare*)

**Figure 9** Extensive growth of *Arthrocnemum macrostachyum* on an upper tidal flat in Albania. This saltmarsh grades into a pine-covered sand dune

Atlantic type of Adam (1990). On the Pacific coast of North America the Saltgrass (*Distichlis spicata*) helps to define the vegetation. From California southwards the California Cord-grass (*Spartina foliosa*) is native as a low-marsh species. Throughout two major embayments, the Gulf of California and in

Washington State, introduced *Spartina* spp. are increasingly invading the intertidal areas, leading to concern and control (Chapter 9).

For more information see *Saltmarshes and Salt Deserts of the World* (Chapman 1974), which provides a description of the vegetation worldwide. In tropical and subtropical zones (roughly between the Tropic of Cancer and the Tropic of Capricorn) mangroves replace saltmarshes. These are not dealt with in this book, but a map showing their distribution is available on the World Conservation Monitoring Centre Web Site (see http://www.unep-wcmc.org/marine/data/coral_mangrove/marine_maps_main.html).

# Chapter 2
# Human Influences

## Traditional Uses, Management and Destruction

## 2.1 Introduction

Despite the apparent and perceived 'naturalness' of the studied systems, few saltmarshes are entirely free from human influence. For centuries, grazing by domestic animals and haymaking took place in Europe and North America. The erection of embankments helped extend the period when grazing could take place and facilitate haymaking. In the early days, these were no more than low earth banks. With improvement in engineering techniques, permanent exclusion of the tide became possible. Some of the subsequent uses have modified the original marsh helping to create new areas with different, but valuable 'semi-natural' assets. Other activities are more destructive. Overall, permanent enclosure removes the natural ability of the saltmarsh to respond to the forces of wind, waves and tidal actions, especially during storms. The description that follows provides a summary of the progressively more destructive changes brought about by human activities and the implications for their nature conservation values.

## 2.2 Traditional Management

Grazing and haymaking are traditional forms of management. Grazing is by far the most widespread activity, though haymaking still occurs in a few locations. Both activities can change the nature of the saltmarsh notably in relation to its component species. Other forms of management include *Salicornia europaea* (Samphire) gathering, turf-cutting and at the margins of saltmarsh, reed beds provide roofing material (thatch). These activities modify the saltmarsh habitat rather than destroying it.

### 2.1.1 *Grazing*

Grazing naturally occurs on saltmarshes involving a wide variety of species. Wintering ducks and geese together with other herbivorous animals such as hares and deer probably relied on these open areas long before human occupation began.

**Figure 10** The effect of grazing – on the right, there is a low-growing grassy sward; on the left taller Sea Aster (*Aster tripolium*) helps provide denser vegetation. In the case of saltmarsh shown here from Bridgwater Bay, Somerset, UK the former provides valuable grazing for ducks and geese, the latter shelter for nesting birds and shelter and food for invertebrates

A variety of ducks and geese, such as wintering Wigeon (*Anas penelope*) in Western Europe, Brent geese (*Branta bernicla*) in the southern North Sea and breeding Lesser Snow geese in Canada continue to graze saltmarshes. Archaeological evidence from the intertidal Severn Estuary shows that prehistoric humans herded animals in these areas (Bell et al. 2000).

Domestic animals can improve the palatability of the swards, as grasses such as *Puccinellia maritima* and *Festuca rubra* are favoured over coarser herbs and shrubs including *Atriplex portulacoides*. The density, period of stocking and type of animal all influence the type of vegetation that develops. At high densities, sheep grazing can create a flat 'bowling green' turf with little structural diversity. Cattle do not graze the sward so tightly, leaving patches of denser vegetation interspersed with low growing areas where grasses are dominant. Visually, the difference between grazed and ungrazed saltmarsh can be dramatic (Figure 10).

## 2.1.2 Reed Cutting

Common Reed (*Phragmites australis*) is a plant, which grows towards the limits of tidal influence, especially where freshwater flows onto the marsh. It is tolerant of brackish, rather than full seawater. It can form extensive areas along the margins of unenclosed saltmarsh and at the upper limits of some estuaries.

The plant has a variety of uses. Prehistoric people along the shores of the Severn Estuary, in the south-west of the UK may have burnt reed to aid herding animals on saltmarsh (Bell et al. 2000). Grazed by domestic stock when young, it also provides fencing (partition fences), coarse mats, baskets, sandals, etc. The straight hollow stems, when cut and dried in autumn provided arrow shafts for American Indians (Duke 1983). However, its principal use is as thatch for roofing. Cutting the reed on a two or more yearly cycle helps maintain the reed at an early stage of its succession. This prevents the development of shrubs and ultimately fen woodland.

### 2.1.3 Samphire Gathering

*Salicornia europaea* (Samphire) is one of the first colonisers of tidal flats. It is a salt-tolerant succulent plant. Brent geese and other herbivores graze it. It has also been gathered and used as a vegetable in salads, or boiled. In developing countries, it is sometimes cultivated as an aid to improving the economic conditions of some coastal communities.

There is no evidence to suggest that Samphire gathering has any significant impact on the development of saltmarsh or its conservation value. However, it represents a human use that has the potential to modify the vegetation succession.

### 2.1.4 Saltmarsh 'Haying'

Haymaking is another human activity that appears to have been widely practised in Europe. It probably continued to take place in the Baltic up to about the 1960s (Dijkema 1990). The first European settlers exported the practice to the eastern USA. Here extensive *Spartina* spp. dominated marshes were a significant source of fodder for their animals (Daiber 1986). In the nineteenth century, saltmarshes in Nova Scotia were dyked in order to grow hay essential to maintain the horse-powered logging and mining industry. By the early 1900s, large tracts of coastal marshes were devoted to this activity. By the late 1930s, the hay market had all but ceased because of fossil-fuel technology (Nova Scotia Museum of Natural History, undated). Today the practice appears to have all but disappeared in both Europe and North America, except for a few isolated places such as New England where farmers have been 'haying' some saltmarshes for over 300 years (Ludlam et al. 2002). It still takes place on a large scale (over 400 ha regularly) throughout Plum Island Sound, northeastern Massachusetts (Williams et al. 2001).

## 2.2 Excavation

Excavating the surface of the marsh is an activity that has taken place for a variety of purposes. These lie somewhere in between the traditional use of the saltmarsh, where tidal inundation continues but where there is little or no modification of the surface and enclosure. As such, there are differing impacts on the saltmarsh as summarised below.

### 2.2.1 Turf Cutting

Heavily grazed saltmarshes can provide turf suitable for lawns, cricket pitches and bowling greens. The extent of this use is uncertain though a combination of fertiliser application, reseeding and intensive sheep grazing produced very short, dense matted turf in parts of north-west England. Once established the turf was periodically cut and sold commercially (Gray 1972). In the Netherlands, Germany and Denmark, turf-cutting provided material for covering and reinforcing sea walls (Kamps 1962). Today the activity continues on a local scale for dyke repairs. However, in Morecambe Bay, a nature conservation site in north-west England, it is one of a number of 'operations' requiring consent from English Nature*.

### 2.2.2 Sediment Extraction

Extraction of saltmarsh sediments, including clay was used to build sea walls in the Wadden Sea (Beeftink 1977a), the Wash, eastern England (French 1997, pp. 62–63) and for brick-making in the Medway in south-east England (French 1997, pp. 95–96). These more dramatic causes of habitat loss create low-lying surfaces where recolonisation takes place. The vegetation can take many years to recover, perhaps showing effects over decades (Beeftink 1977b). In the German Wadden Sea extraction of clay from pits, helps regenerate vegetation succession (including the development of natural creeks) in created, artificially drained saltmarsh (Dijkema pers.com.).

## 2.3 Summer Dykes, Grazing Marsh, Salinas and Rice Fields

Summer dykes and embankments to exclude the tide have progressively greater impact on saltmarshes. The former remove the saltmarsh from tidal influence for only part of the year, usually in the summer. The saltmarsh community remains, albeit with a modified tidal regime. The latter permanently encloses the saltmarsh completely removing it from tidal inundation. This form of enclosure, effectively results in 'land claim' and the creation of land for agricultural or other uses. It affects saltmarshes throughout the world.

---

*English Nature is the Statutory Body responsible for promoting the conservation of wildlife, geology and wild places in England. It is a UK Government agency set up by the Environment Protection Act 1990 and funded by the Department of Environment, Food and Rural Affairs. In 2006, it became part of the wider 'Natural England' with its amalgamation with the Countryside Agency and the Rural Development Service (see http://www.naturalengland.org.uk/).

With permanent tidal exclusion, drainage and water table management are possible. The subsequent use of the land can include grazing, conversion to salinas, use for rice cultivation and with intensive drainage, cultivation with agricultural crops. Each of these creates a different nature conservation interest. This section briefly describes the history of embankment and the type of nature conservation interest surviving under the different types of management. The last of these developments, use for intensive agriculture, ensures the loss of all the saltmarsh features, with little or no replacement of the nature conservation interest. This is included in Section 2.4 dealing with 'land claim'.

## 2.3.1 Summer Dykes

Summer dykes (summerdikes; summerpolders), as the name implies, eliminate tidal flows and reduce sedimentation on saltmarshes, extending the period of grazing or other agricultural use, such as haymaking during the summer. They accompanied the colonisation of the lowlands of the North Sea, for example in the German Wadden Sea (Ahlhorn & Kunz 2002). Today some 1,200 ha in the Netherlands and 2,100 ha in north-west Germany of such partially enclosed saltmarsh remain (Bakker et al. 2002).

Summer dykes are today, more often associated with permanent embankments built to exclude the tide from the saltmarsh. In the Wadden Sea, they form part of the sea defence (Section 2.4.3).

## 2.3.2 Grazing Marsh

Grazing marsh is a habitat type recognised from Great Britain and found mostly in south-east England. It develops following the exclusion of the tide by the embankment of saltmarsh. The important aspect of the habitat is that although some artificial drainage may occur, the original surface of the saltmarsh remains virtually intact. This provides for the development of a suite of topographical features that help to define the habitat. These include remnant creeks (fleets), grassland and transitions between brackish water and fresh water. Gray (1977, p. 257) and Doody (2001, Chapter 11) provide a more detailed description of these features and their nature conservation value. Section 6.4.4 looks at the nature conservation value associated with this habitat in the relation to policies designed to re-create saltmarsh habitat, through realignment.

Grazing marsh thus forms the first of a series of habitats derived from saltmarsh, where modification rather than destruction of the nature conservation value has taken place. They cause progressively greater change to the 'natural' forms, whilst retaining some features of the original habitat (Figure 11).

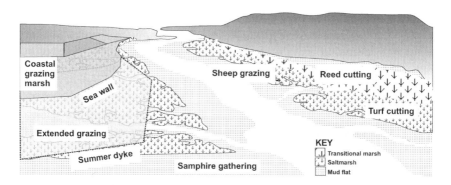

**Figure 11** Some modifications made to the saltmarsh surface by human use, changing, rather than destroying nature conservation values

The next stages in modification result in much greater change and a loss in most of the features associated with saltmarsh.

### 2.3.3 Salinas

A special type of enclosed wetland occurs in warmer locations such as the Mediterranean. Here enclosure of tidal lagoons, which may or may not have saltmarsh, creates saline pools (salinas) for salt production. These may be relatively small 'artisan' salinas (including natural salt lakes) or large industrial complexes. As with coastal grazing marsh, the modified habitat can support important wildlife, though this is to some extent dependent on the intensity of the salt production.

Examples of these habitats include the smaller traditional salinas, largely derived from saltmarsh, such as those on the Algarve (Figure 12), western France and the industrial salinas of the Camargue, which occupy approximately 11,000 ha. The former, depending on the intensity of use (some are abandoned), support salt-tolerant plants such as *Salicornia* spp. and on the drier land Tamarisk (*Tamarisk gallica*). Animals include breeding birds such as Avocet (*Recurvirostra avosetta*) and Shelduck (*Tadorna tadorna*). These species along with several others are especially sensitive to water levels and management can have important consequences for successful breeding (Sadoul et al. 1998). For some birds, such as the Little Tern (*Sterna albifrons*), salinas have become a less favoured but relatively more important nesting habitat in some parts of Portugal, as nesting beaches are increasingly disturbed by tourists (Catry et al. 2004).

Salinas are also important feeding areas not only for birds breeding nearby, but also for those on migration. Water depth is crucial in determining which birds feed where. The Flamingo (*Phoenicopterus ruber roseus*), is often the dominant species and can number up to 10,000 individuals. With decreasing water depth species such as the

**Figure 12** Traditional salinas in the Algarve, Portugal. Note the banks enclosing the pools, which provide nesting sites for a variety of birds. These may, in older and abandoned salinas support interesting plant communities, including species more typical of the surrounding saltmarshes (e.g. *Arthrocnemum macrostachyum*)

Avocet, Black-winged Stilt (*Himantopus himantopus*) are some of the characteristic and more easily recognised species. The shorter-legged Little-ringed Plover (*Charadrius dubius*) feed at the margins of the open water. For more information see Sadoul et al. (1998), Walmsley (1999) and Doody (2001, Chapter 11).

## 2.3.4 Rice Cultivation

Rice cultivation is another widespread activity, which utilises enclosed tidal land throughout the world. It is particularly important in warmer latitudes such as the Carolinas in the USA, parts of the Mediterranean, India and Malaysia. In many of these areas, this use provides important opportunities for intensive and highly productive rice cultivation as in southern Carolina in the USA (Trinkley & Fick 2003). As with salinas under certain circumstances even industrial scale rice cultivation, as in the Ebro Delta, Spain can provide valuable areas for wildlife, especially birds (Ibáñez 1999).

In the Mediterranean, some rice fields provide an alternative habitat for a range of birds threatened by the major loss of coastal wetlands that has taken place there (Fasola & Ruiz 1996). This is especially important for breeding herons in sites such as the Po Delta in north-east Italy, the Rhone Delta in France, the Axios Delta in Greece, and the Ebro Delta in Spain (Fasola et al. 1996). These areas also provide food for migrating birds. This varies seasonally, coinciding with periods of passage. However, agricultural practices may change, causing prey availability and abundance to decline at key stages in the migration cycle. This could result in serious impacts on local migratory water birds (Marques & Vicente 1999).

## 2.4 Enclosure and Habitat Loss – 'Land Claim'

Enclosure of mature saltmarsh, followed by conversion to intensive agriculture (principally arable cultivation and intensive rice fields) usually involves change to the marsh topography. Notwithstanding the value of rice fields for some birds, as identified above, these are generally of a much lower nature conservation interest than the habitats they replace. Enclosure, resulting in the total exclusion of the tide, normally causes a major loss of coastal habitat including saltmarsh and transitions to brackish and freshwater communities.

The use following enclosure has important implications for habitat re-creation (Chapter 6). Building roads or other infrastructure, or the erection of a sea wall followed by infilling, not only destroys the saltmarsh surface, but also leaves few opportunities for restoration to tidal land. Enclosure that allows the surface to remain at or near its original level has the potential for restoration. These activities (summarised in Figure 13) occur in many parts of the world become progressively more difficult to reverse.

Note that excluding the tide prevents the deposition of new sediment behind the sea wall. This, together with compaction due to drainage, results in lowering of the land surface. Sedimentation continues seaward of the sea wall, further increasing the difference between the two. This has important implications for the sustainability to the defences especially on coasts where sea levels are rising relative to the land (see Chapter 10).

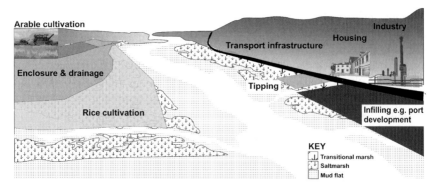

**Figure 13** The main factors causing saltmarsh loss. Those on the left are reversible and habitat can be re-created; those on the right are, for the most part, irreversible

## 2.4.1 Saltmarshes, other Coastal Wetlands and Mosquitoes

Coastal wetlands (saltmarshes and fens) harbour many animals, including mosquitoes that are the vectors of disease, amongst which malaria is the most widespread. Deforestation and erosion of soils in the hinterland help to create vast coastal wetlands, such as those prevalent in the deltas of the Mediterranean. In the eyes of the local population these areas became 'wastelands' with the saltmarshes an important component of the 'mosquito and midge infested swamp'. In some parts of the world, people avoided the problem by settling inland away from these areas. In the Rhone Delta (Camargue), for example, the ancient towns are some distance inland. In other areas where people had already settled, the landscape changed around them. The Ombrone Delta lagoon (on the west coast of Italy) silted up because of deforestation inland. Together with the effects of malaria, the area became uninhabitable, causing abandonment of the harbour around 50 BC (Grove & Rackham 2001).

The same combination of factors may have helped cause the collapse of Greek and Roman civilisations. The process has been described as follows: deforestation results in erosion of soils in the hinterland. Rivers deliver the increasing amounts of sediment to the sea, resulting in the development of coastal lagoons and other wetlands. Mosquitoes breed and infect the local population with malaria. The coastal towns become depopulated and unsustainable (Jones 1907). This is probably only part of the story, nevertheless the effects and nuisance of mosquitoes have spurred on the destruction of saltmarshes and other coastal wetlands not only in Europe, but also in North America (Teal & Teal 1969).

Efforts to control malaria continued in Europe into the middle of the twentieth century. For example, in Albania during the communist period presided over by Enver Hoxha from 1944–1985, drainage of these areas became part of a mostly successful attempt to increase agricultural production and control mosquitoes (Figure 14). Today all the European countries, except Turkey, are free from the disease (World Health Organisation web site, see http://www.euro.who.int/malaria).

## 2.4.2 Enclosure and Drainage of Coastal Wetlands for Agriculture

With or without the incentive of malaria control, drainage of wetlands around estuaries, deltas and other coastal wetlands has taken place for centuries and probably dates back before Roman times. Expanding deltas in the Mediterranean became important places for agriculture, including rice cultivation (as in the Ebro Delta, Spain). Significant areas of coastal land became intensively cultivated land in the Netherlands, Denmark and in England.

The process of conversion varies between different countries. In eastern England, drainage of wetlands (freshwater to brackish water swamps) in the Fenland Basin probably occurred at least, from Roman times, though on a small

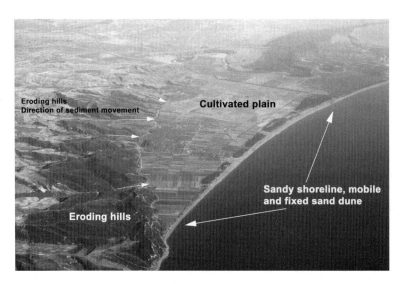

**Figure 14** Coastal plain of northern Albania. A large expanse of sediments derived from erosion in the hills. The arrows show the direction of sediment movement from the hinterland. Much of the area behind the coastal dunes is drained swampland. Note the fringing sand dunes and the absence of towns (Photograph from 1993)

scale. In the 1600s a Dutch engineer, Sir Cornelius Vermuyden, introduced the Dutch methods of enclosure to Britain, and made the first important attempts to drain the area (Figure 15).

Nearer to the coast, successive smaller-scale enclosures in The Wash helped to create new agricultural land. Here a 'counter dyke' prevented the sea from flooding the saltmarsh allowing the excavation of material from a 'borrow dyke', which was used to construct the new earth embankment. The whole process relied on there being sufficient saltmarsh to build both the counter dyke and the embankment, thus the extent of new land was dependent on the rate and extent of saltmarsh development.

Taken together the total area of land drained, claimed and enclosed for agricultural use, amounts to over 130,000 ha. The saltmarsh alone, through piecemeal and cumulative enclosure, provided nearly 30,000 ha by the early 1950s (Dalby 1957). Enclosure continued, with relatively substantial areas of land claim taking place between 1953 and 1983 (Figure 16).

Similar enclosures occurred in the Dee Estuary, in the UK (Pye 1996), the Lower Elbe in Germany (Garniel & Mierwald 1996) and elsewhere in north-west Europe including France and the Wadden Sea (Allen 1997). In the USA, impoundments helped create land for haymaking and other agricultural uses, such as in Louisiana, beginning in the 1700s. A further tranche, promoted by the US Department of Agriculture, began in the early 1900s on the east coast and in Louisiana (Warren 1911). These enclosures are especially important to later discussions, as they present some of the best opportunities for restoration (Chapter 6).

## 2.4 Enclosure and Habitat Loss – 'Land Claim'

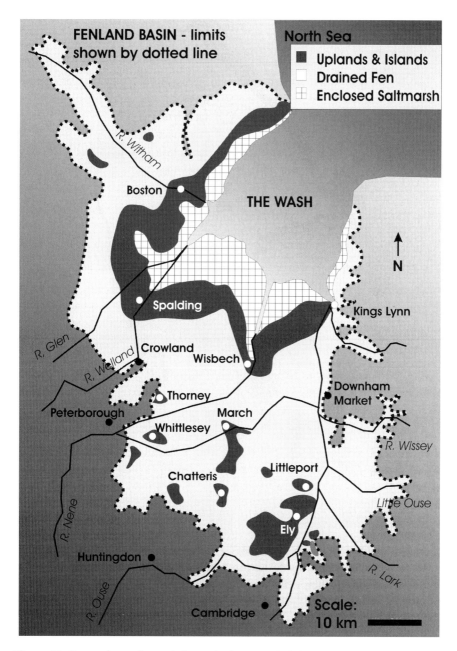

**Figure 15** Progressive and cumulative embankment and drainage in the Fenland Basin, in relation to the saltmarshes of the Wash, eastern England. Redrawn from interpretative material at the Flag Fen Bronze Age Centre, Peterborough, UK http://www.flagfen.com/. See also Brew & Williams (2002) for details of the changes in the extent of the shoreline before embankment. Note Joan Blaeu's map of 1643 also shows the extent of the fens in relation to the 'uplands' particularly well

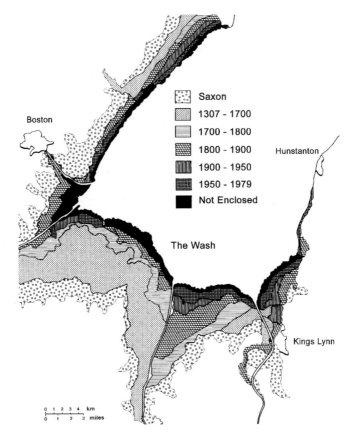

**Figure 16** Sequential enclosure of saltmarsh around the Wash, Lincolnshire, eastern England (After Doody & Barnet 1987)

## 2.4.3 'Warping' and Sediment Fields

In the Wadden Sea (southern North Sea) techniques that are more aggressive were employed, which extended the saltmarsh beyond the normal limits imposed by sediment availability, tides and waves. The process, involved creating a system of 'sediment fields' made of brushwood groynes and low earth embankments intersected by ditches dug in the flats (Figure 17). The increase in sedimentation 'warping' resulting from the enclosed tidal flats with their restricted opening and planting saltmarsh plants such as *Spartina anglica*, helped to promote the establishment of vegetation. The labour-intensive process all but ceased in many areas, from the 1940s (Beeftink 1977a). Sediment fields continue to perform an important sea-defence function by promoting the development of a pioneer zone in front of an eroding cliffed saltmarsh. The brushwood fences aid the processes of sedimentation and plant succession.

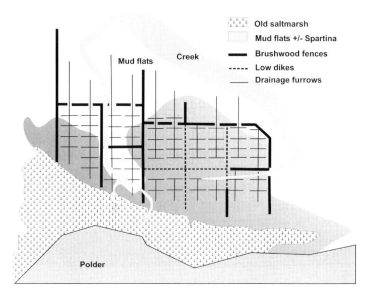

**Figure 17** Sediment fields helping to create saltmarsh in the Netherlands and 'protect' the mature saltmarsh and enclosed 'polder' (after Beeftink 1997a)

Planting *Spartina anglica* to aid the process of colonisation and subsequent enclosure and for coastal protection began in the UK (Gray & Raybould 1997), became common throughout the world (Ranwell 1967), in Ireland (Nairn 1986), New Zealand (Allan 1930) and China (Chung 1990). The issues surrounding *Spartina* spp. its value or otherwise, including the effects on tidal flats, invertebrates, wintering birds and native saltmarsh plants is considered later (Chapter 9).

## 2.4.4  Enclosure for Infrastructure

Enclosure within estuaries and the creation of new land through infilling is a worldwide activity. In the Netherlands the area called the Zuiderzee, a saltwater inlet of the North Sea became an area of more than 150,000 ha of usable land by 1932 and a freshwater lake, the Ijsselmeer. Following the storm surge of 1953, building structures associated with the Delta Project resulted in the loss of large areas of intertidal land (d'Angremond 2003).

In the USA, the major causes of coastal wetland loss are urban development and encroachment by coastal waters caused by impoundments, dredging projects and rising sea level. Losses are concentrated in highly developed areas, such as the Gulf of Mexico and Chesapeake Bay (Johnston et al. 1995). Because of the scale of the enclosures, it is difficult to identify losses associated with saltmarsh (and mangroves) specifically, as opposed to tidal and freshwater wetlands more generally.

**Figure 18** A scene from the 1980s, early 1990s in Great Britain illustrative of the 'view' of some local authorities and other organisations, of the 'value' of saltmarshes and tidal areas

The situation is similar when looking at the extent of the losses of saltmarsh on a worldwide basis. Most references are for tidal wetlands and there is no distinction made between saltmarshes, tidal lagoons, sand or mudflats. However, the losses associated with these enclosures are significant and there are major areas of 'reclaimed'* tidal land around the world (Wong 2005).

In addition to the major areas of enclosure, there is also a much more widespread pattern of piecemeal loss. This is often associated with a prevailing view that saltmarshes and their associated habitats are little more than 'wastelands'. A study of estuaries in Great Britain, for example, showed that in 1989 of 123 cases of estuary land claim, more than 50% affected intertidal or subtidal habitats. Of the approximately 1,000 ha of land affected, over 60% of the intertidal areas suffered from rubbish and spoil disposal (Davidson et al. 1991; Figure 18).

## 2.4.5 Summary of Enclosure

A key issue for the management, and more importantly restoration of saltmarshes lost through enclosure, is the after use of the land. Enclosure, for use in crop production, requires the removal of the saltmarsh from tidal influence. There need be little or no modification of the surface except for the infilling of tidal creeks and

---

*The use of the word 'reclamation' in the context of this book is inappropriate as the saltmarsh is not reclaimed (retrieved or recovered as defined by the Compact Oxford English Dictionary) but destroyed.

removal of other surface irregularities. Drainage is important and once the salt has leached out of the soil, planting takes place. The saltmarsh can, within just a few years become high-quality agricultural land. The key point here is that reversal of the process is possible.

By contrast, saltmarshes that have been dredged and infilled, developed for industry, airports or housing are lost forever. Large enclosures such as the significant areas of the land 'won' from the sea in the Netherlands are unlikely to become tidal again, because of their high social and economic values.

## 2.5 Other Influences

Many other factors influence saltmarshes and their development. Some, such as recreational activity have relatively little adverse impact. Bird watching, walking, boat access and wildfowling, may have localised effects on the saltmarsh, but are more likely to be positive assets when looked at in relation to the value of the habitat. Others, such as oil and litter pollution can be more damaging. Salt production, aquaculture, insect control (especially for the mosquito), tidal power, tidal barriers, water storage and introduced species all have direct and/or indirect effects on saltmarsh. The way in, which these activities influence the conservation, management and restoration of saltmarshes is discussed later.

The final and in some ways most significant human influence on saltmarshes is global warming and the effect this is thought to have on sea levels. Although some argue there is no direct relationship, sea-level change has an important influence on the direction of saltmarsh development. The combined effect of saltmarsh loss as a direct result of human activities and sea-level rise has a significant impact on the sequence of tidal habitats including saltmarshes (Section 3.5.2, the 'saltmarsh squeeze').

The deliberate reversal of enclosure in estuaries is known as 'managed realignment' in the UK and 're-integration with the sea' in Germany. This is a process, which includes flooding of previously defended land, in an attempt to encourage the re-creation of tidal land (Leggett et al. 2004). It is normally associated with sea-defence issues, but also has important implications for nature conservation. It forms part of the discussion in Chapter 6, on saltmarsh restoration.

# Chapter 3
# Nature Conservation

## Changing Perceptions and Policy

## 3.1 Introduction

Chapter 2 describes ways in which the increasing human population has altered the natural coastal environment. The loss of species and habitat due to human exploitation led to the development of the nature conservation movement, not just for coastal areas but more generally. Initially the conservation of species and habitats was the concern of scientists, naturalists and others interested in nature. Conservation often meant '**preservation**' of the environment and natural resources. The cornerstone of the nature conservation effort was the selection of 'protected' sites, using criteria designed to identify the most valuable natural or semi-natural assets, normally in terms of habitats and species. Once established and boundaries drawn, the sites became 'sacrosanct'. Their future conservation rested on the prevention of loss and damage from harmful human developments. Where change occurred, which resulted in the original nature conservation values deteriorating, **management** formed a key part of the conservation strategy.

In the past, developers, politicians and the public were at best indifferent to the loss of the habitats and species or at worst actively engaged in their destruction. This situation changed as the interest in the principles of nature conservation developed and perhaps more importantly, with the recognition of the 'value' of some of these habitats and ecosystems to the socio-economic well-being of human society.

Undoubtedly, some of the greatest losses in absolute terms have been in the coastal wetlands of the USA. Here, and in Europe, recognition of the impact on the coastal environment generally led to policy changes affecting the conservation, management and restoration of coastal areas, generally. In the UK, saltmarshes have been more obviously associated with policy change and restoration initiatives. The purpose of this chapter is to describe some of the ways in which a change in the perception of the value of coastal habitats, especially of saltmarshes, has influenced conservation and management policy. In particular, it emphasises the change in understanding of the wider implications of habitat destruction.

## 3.2 The USA

In the USA, the destruction of estuary ecosystems (including saltmarsh) was on a truly monumental scale. Estimates of this loss show that 'from colonial times up to 1990, over 55 million acres of wetlands in the coastal mainland States were degraded or destroyed. This accounts for more than 50% of the total wetlands losses throughout the nation.' (Estuary Habitat Restoration Partnership Act 1999).

In answer to the question 'How Much Estuary Habitat Have We Lost?' the Restore America's Estuaries Web Site (see www.estuaries.org), lists the following more detailed information for the losses over the last 100 years (Table 2).

Losses in each estuary often involve an accumulation of small developments and enclosures, a 'death by a thousand cuts', as well as much larger developments. These have taken place with little regard for the many economic values and quality-of-life benefits, derived from coastal wetlands. Population growth in coastal watersheds; dredging, draining, bulldozing and paving; pollution; dams and sewage discharges, are among the impacts that have led to the extensive loss and continuing destruction of estuary habitat.

Recognising the economic and social significance of estuaries (including their saltmarshes), the USA enacted the Estuary Restoration Act (ERA), which was signed into law in November 2000. It makes restoring the nation's estuaries a national priority. [The National Oceanic and Atmospheric Administration (NOAA) have an excellent web site providing details of the Act and its implementation

**Table 2** Estuarine habitat loss nationwide in the USA

| | |
|---|---|
| Puget Sound | 73% of the original saltmarshes have been destroyed; |
| Narragansett Bay | 70% of saltmarshes are being cut off from full tidal flow and 50% have been filled |
| San Francisco Bay | 95% of its original wetlands have been destroyed; only 300 of the original 6,000 miles of stream habitat in the Central Valley support spawning salmon |
| Galveston Bay | 85% of seagrass meadows |
| Louisiana estuaries | continue to lose 25,000 acres annually of coastal marshes, roughly the size of Washington, DC |
| Hudson Raritan Estuary | 75% of the original tidal marshes have been destroyed in both New York and New Jersey, and 99% of New York's fresh wetlands are gone |
| Chesapeake Bay | 90% of seagrass meadows were destroyed by 1990; in 30 years (1959–1989), oyster harvest fell from 25 million pounds to 1 million pounds |
| Long Island Sound | More than 40% of the Sound's tidal wetlands have been destroyed |
| Gulf of Maine | Since 1975, developed land in the lower watershed has doubled |
| North Carolina Estuaries | The state has lost more wetlands than any other from 1973 to 1983 |
| Tampa Bay | 80% of seagrass meadows destroyed |

(see http://era.noaa.gov/).] As part of the Act, NOAA is required to develop and maintain a database of estuary restoration projects. A search of this inventory (see https://neri.noaa.gov/query_main.html) reveals 252 estuary restoration projects throughout the USA, of which 60% are devoted solely to the restoration of saltmarsh. The inventory:

- Provides information to improve restoration methods;
- Provides information for reports transmitted to Congress (Section 108, b);
- Tracks the acres of habitat restored towards the one million acre goal of the Act.

Many of these projects are community-based and involve the planting of saltmarsh plants to encourage the establishment of saltmarsh as part of a shoreline stabilisation and fish production programme. For those interested in the reasons for restoration and the techniques employed there are detailed reports available from the inventory, of the costs and methods employed.

## 3.3 The Changing Scene in Europe

In 1984, an inventory of European saltmarshes heralded an increase of interest in them, as an important coastal habitat (Dijkema 1984). This was to be part of a review of coasts and coastal habitats more generally. However, funding from the Council of Europe, which commissioned the saltmarsh study, ceased and for several years interest in individual habitats waned, at this level. Extracting information specifically on saltmarshes from other more recent European wetland inventories is difficult. Such information as there is relates to wetlands as a whole, and these data are incomplete. Difficulties of definition also mean that little meaningful information is available for saltmarsh. For example, a summary of wetland loss based on the 'best available' information, only quotes two relevant pieces of information, losses of 'Coastal wetland' in Albania, and 'Saltmarsh' in the UK (Moser 2000)

However, despite the paucity of information, individual countries have seen major changes in policy. The UK provides an example of the way the process advanced.

## 3.4 Habitat Loss and Nature Conservation in the UK

The nature of enclosure, and subsequent use of saltmarsh described in Section 2.4, represents a major cause of habitat loss. The value of the new land, and economic development opportunities, meant that few expressed concerns about the long-term implications of the enclosures, or ignored them altogether. As a result, the loss of tidal land, including saltmarsh, continued unabated throughout the twentieth century. A review of the situation in Great Britain in the late 1990s showed a cumulative loss of some 25% of the natural intertidal areas on 155 estuaries over

## 3.4 Habitat Loss and Nature Conservation in the UK

**Table 3** Some key infrastructure projects with significant estuary habitat loss and knock-on impacts on waterfowl. Note in the UK, of the infrastructure projects only the Cardiff Bay Barrage was completed. The reasons for the failure of the other proposals rested on financial considerations, in a changing economic climate. Further information from the *Journal of Aquatic Ecology* (1978, 12/3–4), which has 16 papers covering the Delta works in the Netherlands and the water storage and energy generation schemes in the UK

| Development | Site | Main issues | References |
|---|---|---|---|
| Energy generation | The Severn Estuary | Loss and change in habitat & estuary status; loss of extreme tidal range | Various proposals 1849–1989 and 2006 (Wikipedia 2007) |
| Water storage (freshwater) | The Wash | Loss of habitats and winter feeding areas for birds | Ruxton & Baker (1978) |
|  | Morecambe Bay | Loss of habitats and winter feeding areas for waterfowl |  |
| Airport construction | Maplin Sands | Loss of habitat and winter feeding areas for | Department of Trade (1974) |
| Port and harbour development | Tees Estuary (Seal Sands) | Loss of habitat for winter feeding birds and seals, 'jobs versus birds' | Cleveland Structure Plan, Examination in Public 1975 |
| Enclosure for agriculture | Extension of the historical 'reclamations' | Cumulative loss of foreshore habitats (saltmarsh and mudflats) and reduced feeding areas for wintering waterfowl | Various sources |
| Economic regeneration (amenity barrage) | Cardiff Bay | Loss of estuary, displacement of wintering waders | Cardiff Bay Barrage Bill and Select Committees of the House of Commons and the House of Lords in 1990/1 |

the previous 100 years or so. In 1989, there were 123 cases of land claim affecting 45 of these sites (Davidson et al. 1991).

The nature conservation movement, recognising the potential harmful effect of further, often cumulative developments on the wildlife value of the affected areas battled, against them for two decades and more. During the latter part of the twentieth century, some of the more significant proposals were for energy generation, water storage, infrastructure development and agriculture (Table 3), all affecting saltmarshes to a greater or lesser extent.

Throughout the 1970s and 1980s the conservation case against further enclosure and development on the estuaries of the UK rested on four principal arguments:

1. The inherent value of the areas affected;
2. Direct loss of habitats and their associated plants and animals in relation to features of national and international importance;

3. The indirect impact on the feeding areas of wintering bird and their ability to sustain numbers at national and internationally recognised levels;
4. The implications for the wider conservation of migratory species using a network of sites, including those in the UK.

Most of the early cases became simple choices, as far as those promoting the schemes were concerned, socio-economic progress or nature conservation, (often expressed as jobs versus birds). Most of the cases, found to be technically feasible and affordable were lost, with the nature conservation arguments considered to be of secondary importance.

These, and similar cases helped reinforce the belief amongst conservationists, at least, that there was a need to launch campaigns to raise awareness of the problems. These included the English Nature, 'Campaign for a Living Coast', which resulted partly from the evidence presented in the 'Estuaries Review' (Davidson et al. 1991) and the Royal Society for the Protection of Birds 'Save our Shorebirds' campaign in 1993. Three case studies illustrate the way the conservation argument and the subsequent view of enclosure of tidal saltmarsh have changed in recent years.

## 3.4.1 Saltmarsh Enclosure in the Wash, South-East England

The history of enclosure in the Wash (Figure 16) seemed to suggest that the creation of new land from the sea provided never-ending opportunities for landowners to add to their land. With intent, money and machinery it was possible to create new land by erecting an earth embankment enclosing mature saltmarsh that was 'ripe for reclamation'. In just a few years, this could be cultivated with arable crops (or even daffodils!). The benefit to the landowners was obvious, as they were able to acquire new land at relatively low cost. Subsequent maintenance of the sea wall became the responsibility of the National Rivers Authority (now the Environment Agency for England and Wales).

In the late 1970s and early 1980s the Nature Conservancy Council (NCC) expressed concern about the impact of the enclosures for nature conservation. Although enclosures were often quite small (100–300 ha) their cumulative impact was large. The location, size and dates of enclosure are detailed in *Memoirs of the Geological Survey* (Gallois 1994). In the Wash, as elsewhere in the southern North Sea, embankment of saltmarsh (Figure 19), not only destroyed the habitat in the short term, but also reduced its overall diversity because of the continual removal of the more mature and botanically diverse sections of marsh.

At the same time, although the evidence was not conclusive, the suggestion was that the overall intertidal area was not advancing progressively seawards. Although new marsh developed outside the sea bank, it did so at a rate greater than the seaward movement of low water mark. Thus, as the saltmarsh regenerated outside the sea wall, there appeared to be a consequent loss of sand and mudflats in the

## 3.4 Habitat Loss and Nature Conservation in the UK

**Figure 19** Erection of an earthen embankment, southwest shore of the Wash in the late 1970s

Wash. Since these provided important winter feeding areas for large numbers of wildfowl and waders there appeared to be a potential for significant loss of value in these populations. Based on the evidence, several nature conservation organisations objected at a public inquiry in February 1980, into a further piecemeal saltmarsh enclosure at Gedney Drove End. The Government Inspector upheld a local inquiry into a moratorium on saltmarsh enclosure proposed by the Lincolnshire County Council in their coastal subject plan of 1983.

A conference held in Horncastle, Lincolnshire, England reported the results of a study on the effects of the enclosures between 1971 and 1985 (Doody & Barnet eds. 1987). These showed a loss of 865 ha of marsh but with accretion rates of over 20 mm per annum, new marsh totalling 781 ha developed over the same period (Hill & Randerson 1987; Hill 1988). The report also includes detailed reports on the sediment budget for the site (Dugdale et al. 1987; Evans & Collins 1987) and the potential impacts of sea-level rise (Shennan 1987).

Taken together these results and other papers, helped to confirm several important consequences of saltmarsh enclosure for agricultural use in the Wash, namely:

1. Losses usually involve the more mature high-level plant communities and their associated invertebrates and breeding birds. Whilst new marsh develops to seaward this takes many years to fully replace the biological diversity lost through enclosure;
2. Low Water Mark remains more or less static, or at least does not move seaward at the same rate as the advance of saltmarsh following enclosure. This situation was likely to be aggravated by relative sea-level rise;
3. Reduction of the intertidal area as new marsh, in some cases with an accelerated accretion rate, extends beyond the new sea wall. Continuing saltmarsh enclosure results in a reduction in the area of sand and mud flats available, particularly to feeding wintering ducks and geese.

These impacts led to the conclusion that 'each incursion into the intertidal zone represents a loss of interest and these losses are cumulative.' It was also suggested that 'any further *squeezing* of the intertidal zone' must upset the balance of processes 'essential to the well-being of the system.' (Doody 1987).

The last significant enclosure took place since 1979 along the Freiston Shore, which became a 'managed realignment' site in 2004 (Section 3.5.3). A recent review of the historical changes to the shoreline of the Wash suggests that some sections are accreting whilst others are retreating. Looked at over the period from 1828–1995, there is a general trend showing low water mark moving seawards. However, this situation appears to have changed in recent years, with a landward movement identified along significant stretches of the shore from 1971–1995. Predictions of future change depend on the rate of relative sea-level rise, and the overall progression landward or seaward remains inconclusive (Brew & Williams 2002).

## 3.4.2 Erosion of Essex Saltmarsh, South-East England

Extensive losses of saltmarsh in Essex and north Kent due to enclosure and subsequent use for agriculture, including coastal grazing marsh, were in the order of 4,340 ha, according to studies undertaken to assess the implication of building a third London airport at Maplin Sands (Boorman & Ranwell 1977). Visual evidence suggested that the remaining marshes were suffering further degradation through erosion. The situation at Northey Island is illustrative of the process occurring throughout the saltmarshes on the Essex (and Kent) coasts in south-east England over the last 150 years or so, as tidal land was enclosed, and then abandoned as sea walls failed.

**Northey Island saltmarsh – history**

In 1840, the tithe map shows that most of the island to have been used as pastures (P) or arable land (A) (Figure 20). After a major storm in 1897, large areas (approximately 2/3 of the island) became flooded following breaches to the sea wall, which were not repaired. By 1901, these areas had reverted to saltmarsh (Figure 21).

The general perceived trend in saltmarsh erosion prompted the NCC to undertake a review of change in saltmarshes along the Essex coast. Detailed surveys were undertaken using aerial photographs flown in 1988 and compared with maps prepared from 1973 aerial photographs (Burd 1992).

In 1991, a small 0.8 ha site on the island was the first deliberate attempt at experimental managed realignment in the UK. The engineered realignment was a collaborative effort between the owners of the site, the National Trust, the National Rivers Authority (now Environment Agency for England and Wales) and English Nature (now part of Natural England). A study of plant recolonisation showed that by 1993 a pioneer saltmarsh community had developed. By 1994, there were

## 3.4 Habitat Loss and Nature Conservation in the UK

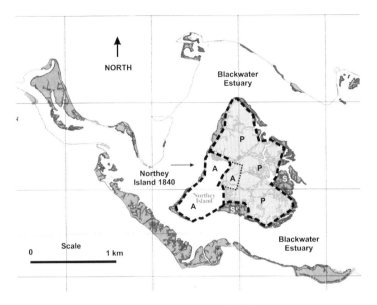

**Figure 20** Northey Island as shown on the Tithe Map of 1840

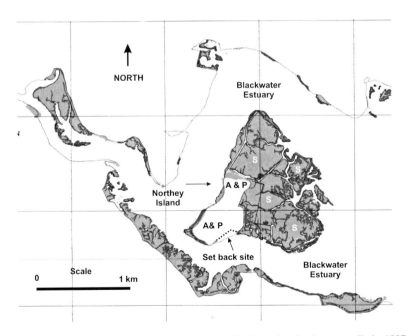

**Figure 21** Gain in saltmarsh (S) following uncontrolled breaches in the sea walls in 1897, at Northey Island in the Blackwater Estuary, Essex. The map shows the areas of saltmarsh loss (dark grey), which occur along the creeks and within the body of the saltmarsh (from Burd 1992)

25 species, including the rare Shrubby Sea-blite (*Suaeda vera*) and other recognisable saltmarsh communities, forming mosaics with the pioneer marsh (Dagley 1995). This experiment marked a significant, if small, step towards developing managed realignment as an alternative to continuing strengthening of existing sea walls.

An update and reassessment of the previous surveys, using a further series of aerial photographs flown in 1998, confirmed the general erosion trend in the saltmarshes of south-east England (Cooper et al. 2000, summarised in Cooper et al. 2001, see also Table 4).

**Table 4** Net loss of saltmarsh in the Essex estuaries in hectares. NB the figures take account of losses due to erosion and enclosure, and gains because of accretion

| Area (ha) | 1973 | 1988 | 1998 | % Loss (1973–1998) |
|---|---|---|---|---|
| Stour | 264 | 148 | 107 | 59 |
| Hamford Water | 876 | 765 | 621 | 29 |
| Colne | 792 | 744 | 695 | 12 |
| Blackwater | 880 | 739 | 684 | 22 |
| Dengie | 474 | 437 | 410 | 13 |
| Crouch | 467 | 347 | 308 | 34 |

The estimated total net loss of saltmarsh over the 25-year period was in excess of 1,000 ha (approximately 25% of the area present in 1973). Enclosure for port and other developments accounted for only 5% of the observed loss, with the rest resulting from erosion at the margins of the saltmarsh, both within the marsh, along creeks and on exposed shores.

Also in Essex, the Dengie Peninsula, the largest single area of saltmarsh in the county, lost approximately 10% of its surface area over a 21-year period (Harmsworth & Long 1986). Evidence of more general habitat loss came from the 'Anglian Sea Defence Management Study' (Sir William Halcrow and Partners Ltd. 1988a, 1988b). It concluded that 70% of the coastline from Humberside to the Thames was retreating. The loss of saltmarsh to erosion appears to be part of a more general loss of intertidal habitat, with the foreshore becoming steeper and narrower. Along the coast of England and Wales, a more recent study showed that 61% of the shoreline had a tendency towards steepening. This appears to be the dominant state of the shorelines of the west, south and east coasts (Taylor et al. 2004).

In the absence of enclosure, saltmarshes build up at the expense of sediment derived from the adjacent mudflat (van de Koppel et al. 2005). The foreshore becomes steeper until wave action initiates erosion. This may result from a storm event at which point the saltmarsh might enter a phase of 'galloping' retreat (Allen pers. com.). This situation will result in a landward movement of the saltmarsh where the topographic conditions allow. Where there is an impediment to landward movement, as with rising ground or in Essex a sea wall, further advance in front of the eroding saltmarsh depends on external forces (e.g. invasion of *Spartina anglica* or an increase in sediment availability) promoting growth. The combined effects of a relative rise in sea level, increased wave activity (possibly associated with global warming) and an embanked landward margin appear to be instrumental in promoting their erosion.

## 3.4.3 Cardiff Bay, South Wales

Discussions took place in the late 1980s and early 1990s to build an amenity barrage across Cardiff Bay as part of the regeneration of a derelict area with 'unsightly' tidal flats and saltmarshes. In the words of the Official Cardiff Bay Website (see http://www.cardiffbay.co.uk/) 'the barrage will eliminate the effect of the tide, which has acted as an inhibitor to development, releasing the potential of the capital city's greatest asset – its waterfront'. Opposition to the proposal rested on the implications of the direct loss of habitat and the conservation of migratory birds using the Severn Estuary, a site of international nature conservation importance. Despite considerable and comprehensive evidence presented to Select Committee hearings of the House of Commons and House of Lords, by the NCC and others, Cardiff Bay was effectively lost with the construction of the barrage in the late 1990s, following the enactment of the Cardiff Bay Barrage Act in 1993.

Although the strong conservation case could not overturn the political pressure for an 'amenity lake', compensation for its loss became an important part of follow up action. In particular the UK Government agreed that development and maintenance of a wetland reserve involving the creation of saline pools, reed beds, and managed grassland on the Gwent Levels, should be made to 'compensate' for the loss of the tidal land in Cardiff Bay, some 15 km or so to the west.

Eventually, the European Commission accepted the proposal. In February 1996, they wrote confirming this to the Royal Society for the Protection of Birds (RSPB), who had lodged a formal complaint in relation to the barrage under the Wild Birds Directive. It stated its acceptance of the UK Government's view that the construction of the barrage could be justified for reasons of overriding socio-economic interest. The letter continued:

> "The UK authorities have also given guarantees concerning the measures to be taken to satisfy the criteria of Article 6(4) of Council Directive 92/43/EEC. The wetland compensation and conservation measures to be taken by the UK authorities include substantial measures to create new wetland habitat and put in place additional management plans for 31 estuaries in the UK."

## 3.5 From Enclosure to Enlightenment and Realignment

The three cases referred to above illustrate different elements in the changing approach to coastal protection from both a nature conservation and engineering perspective in the UK. The situation in the Wash undoubtedly led to recognition that it was no longer feasible to enclose tidal land without permanent loss of intertidal habitat. It also seems that the diminishing scale of saltmarsh that was, 'ripe for reclamation' and the reduction in economic return from the 'new' land, contributed to the cessation of enclosure.

The results of the studies on the loss of saltmarsh further south on the Essex Coast (Figure 22), seemed to suggest that continuing the policy of maintaining the

**Figure 22** Loss of saltmarsh on the Essex coast, south-east England. Note, as erosion proceeds, exposure of previous attempts to stabilise the saltmarsh occurs

existing line of defence would eventually lead to the loss of further substantial areas of saltmarsh.

There followed a review of these losses including those of other coastal habitats and targets for habitat restoration (Pye & French 1992). This helped provide the basis for the establishment of a Biodiversity Action Plan for coastal saltmarsh (see http://www.ukbap.org.uk/UKPlans.aspx?ID=33). This provides for the creation of 40 ha of saltmarsh in each year of the plan to replace the 600 ha lost between 1992 and 1998, in addition to 100 ha to replace the estimated continuing losses. The response to eroding saltmarsh in eastern England mirrors the changing thinking amongst policy makers.

## 3.5.1 'Protecting' Essex Saltmarsh

In Essex, the first response to the eroding saltmarsh in the early 1970s was to use techniques borrowed from the Dutch, to help re-create saltmarsh in badly eroded areas. A series of experimental sites were set up and a variety of techniques used to encourage sediment accretion, such as the polder-like structure at Cudmore Grove, Essex (Figure 23).

Overall, the results of this and other similar experiments in Essex were disappointing (Holder & Burd 1990). Coupled with this the increasing cost of maintaining the existing line of defence to protect areas of limited agricultural

## 3.5 From Enclosure to Enlightenment and Realignment

**Figure 23** Wooden 'groynes' built on the muddy foreshore to form a polder-like structure as part of an attempt to re-create saltmarsh and prevent cliff erosion at Cudmore Grove, Mersea Island, Essex. Monitoring showed that by 1990 (some 14 months after installation) *Spartina anglica*, present when construction took place, was still there. There was also some evidence of accretion in sheltered parts of the polder. However, by 2002 as the above picture shows, there were no signs of saltmarsh development

value, led to a growing recognition that simply 'holding the line' might have further adverse consequences for nature conservation and might not always be cost-effective for sea defence.

### 3.5.2 The UK 'Saltmarsh Squeeze'

The combination of saltmarsh enclosure and a static or receding low water mark in the Wash, coupled with the recognition that this led to a narrowing foreshore helped define 'coastal squeeze' (Doody 2004). The more specific 'saltmarsh squeeze' is the process where removal of saltmarsh from the influence of the sea, through enclosure, causes the direct loss of habitat. In areas with relative sea-level rise the inability of the habitat to 'roll-over' landwards, because of the presence of hard (static) sea defences, results in a 'squeeze' on the saltmarsh (Pethick 2001). Any reduction in sediment availability results in further shortening and narrowing of the beach and a reduced ability for new saltmarsh to develop. Increasing wave attack, caused by climate change and/or sea-level rise, will exacerbate the effect. This in turn not only affected the nature conservation interest of many nationally and internationally important sites, but also left the existing sea walls more prone to attack by waves and storms.

### 3.5.3 Enlightenment

The lack of success in re-creating saltmarsh seawards of eroding shorelines in Essex, the recognition of the processes resulting in 'saltmarsh squeeze' and the relatively low economic value of some of the surrounding land led, to a change in attitude

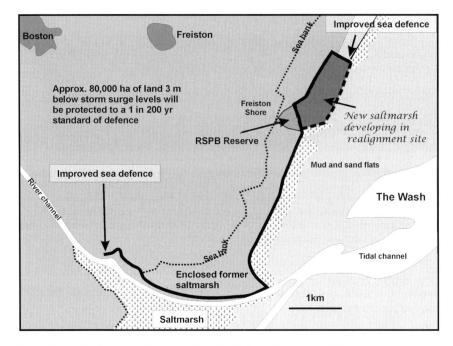

**Figure 24** A 'field scale' realignment along the Freiston Shore on the Wash

towards sea defence. In England, this manifested itself in a series of experimental 'managed realignments' (dealt with in more detail in Chapter 6). Completed in the year 2000 one of these was at Freiston Shore on the Wash (Figure 24). The total area of saltmarsh and mudflat 'reclaimed' by the sea through 'realignment' at this site is 78 ha. It is perhaps appropriate that this area, which was the last enclosure of saltmarsh on the Wash, should be one of the first 'managed realignment' along this stretch of shoreline.

In extreme cases of 'coastal squeeze', the loss of habitat may be such that creating new habitat adjacent to or even away from the existing site may be the only compensation option. The loss of the estuary of Cardiff Bay, South Wales to create an amenity lake as part of a scheme to regenerate the Bay is such a case. The recognition (by the UK Government and the European Commission) that new replacement habitat nearby and improved management of all UK estuaries, was part of a wider compensation mechanism is important. Today, some 5 years later, a new wetland nearby has developed a significant, if different, wildlife value to the original tidal flats (Williams & Phillips 2005). Despite this replacement habitat, others claim that the £200 million Cardiff Bay barrage, provided few real benefits, caused major ecological damage and that the regeneration of that part of Cardiff would have happened without it (Best 2004).

This case is part of a subtle change in policy occurring in the UK. This has moved, in the relatively short space of 20 years, from a situation where coastal habitats in general and saltmarshes in particular, are a positive asset rather than just 'wastelands'. Not only has the enclosure of tidal land for agriculture ceased, but also increasingly bold steps to restore coastal habitats, through the reclamation of tidal land lost to agriculture, have taken place.

## 3.6 The Rest of Europe

In mainland Europe, conservationists made similar arguments to those advanced in the UK. In Germany, the enclosure of tidal marshes had continued up to the 1990s. From the 1960s onwards, the primary objective of embankment had been for sea defences. However, recognition that these had a detrimental impact, especially on the biologically rich areas of the German Wadden Sea, resulted in a decline in these projects. A number of factors came together to persuade the authorities that continuing enclosure was not in the best interest on the environment or for sea defence. These were:

- Objections from environmentalists;
- A reduced need for increases in agricultural production;
- New, more environmentally sound legislation;
- Recognition that sea-level rise was an important issue (Goeldner 1999).

In the Netherlands, more specific nature conservation issues surrounded the consideration of the significance of the loss of saltmarsh, through embankment on goose populations (Dijkema 1987). Whilst the numbers of Dark-bellied Brent Goose (*Branta bernicla bernicla*) did not appear to be directly affected, displaced birds had a lower breeding success and survived less well. There was also evidence to suggest that the loss of a spring-staging site increases population pressure on remaining sites, possibly reducing the fitness of the population as a whole. In addition, the displaced birds invade surrounding fields leading to increasing conflict with agricultural interests (Ganter et al. 1998). As in Germany, this information helped to promote the case for cessation of enclosure.

### 3.6.1 Europe – ICZM, Erosion and All That

Despite the recent lack of information relating to saltmarsh specifically, policy initiatives involving coastal areas have taken place in Europe. Within the European Union (EU), the development of policy in this area probably became more significant in 1981, when a conference on Maritime Peripheral Regions of the Community drew up a European Coastal Charter. Article two of the EC 5th Action Plan 'Towards Sustainability', produced by the Commission of the European Communities in 1992, introduced the concept of sustainable development and included several target areas, of which the coastal zone was one. The stated objective was 'Sustainable development of coastal zones and their resources in accordance with the carrying capacity of coastal environments'. Subsequent action following on from the EU's 5th Environment Action Plan led to an EU Demonstration Programme.

The European Commission operated a Demonstration Programme on Integrated Coastal Zone Management (ICZM), designed around a series of 35 demonstration projects and 6 thematic studies, from 1996 to 1999 (European Commission 1999). Details of this and follow-up initiatives are available on the Commissions' web site (see http://europa.eu.int/comm/environment/iczm/home.htm#zone1). In 2000, based on the results of the Demonstration Programme the Commission adopted two documents:

1. Communication from the Commission to the Council and the European Parliament on 'Integrated Coastal Zone Management: A Strategy for Europe' (COM/00/547 of 17 September 2000), (see http://europa.eu.int/comm/environment/iczm/comm2000.htm);
2. A proposal for a European Parliament and Council Recommendation concerning the implementation of Integrated Coastal Zone Management in Europe (COM/00/545 of 8 September 2000). This recommendation was adopted by Council and Parliament on 30 May 2002.

As part of the further measures promoting Integrated Coastal Zone management, the European Parliament asked the European Commission to undertake a review of coastal erosion policy and practice throughout the EU. At the time, 20% of Europe's coasts appeared to be experiencing severe impacts from coastal erosion. Erosion management to combat this reached an estimated € 3,200 million in 2001. Most of the money provided for ad hoc solutions, many of which failed, or created additional problems nearby. A pan-European study, with the acronym 'eurosion' (see http://www.eurosion.org/) ran from January 2002 to May 2004. The main findings and recommendations from the project are summarised in a brochure 'Living with coastal erosion in Europe: Sediment and Space for Sustainability' (Doody et al. 2004).

### 3.6.2 Accommodating Change – 'Living with the Sea'

The driving force for the EU LIFE project 'Living with the Sea' was 'coastal squeeze' as it affected intertidal Natura 2000 sites. [Natura 2000 is a network of protected sites derived from sites selected as Special Areas of Conservation under the EU 'Habitats Directive' and Special Protection Areas under the EU 'Birds Directive'. For more information, see the European Commission Nature and Biodiversity Home Page (see http://europa.eu.int/comm/environment/nature/home.htm.)] As described above human activities had for many years 'fixed' the coastline with artificial structures to control erosion and where possible created new land from the sea. The study looked at sites in south-east England which, as has been described above, is an area suffering from rising sea levels and a long history of tidal enclosure.

## 3.6 The Rest of Europe

A key question posed by the project was how to deliver 'Habitat Directive compliant flood and coastal defence schemes' (Worrall 2005).

Part of the problem addressed lies in the fact that statutorily designated areas and nature reserves have fixed boundaries. In areas where relative sea level is rising, with or without erosion taking place, 'natural' change may result in habitat loss. In the most dynamic areas, coastal habitats will evolve in such a way as to move beyond the limits of 'protected' sites. Taken together, these factors suggest the need for a wider appreciation of the role of natural processes and sediment dynamics in coastal conservation and management. In this context and given the extent of the accumulated losses of coastal habitats Europe-wide, the restoration, re-creation or creation of coastal saltmarshes must be an essential part of any nature conservation effort.

The rest of this book sets out to provide information on the trends, trade-offs and possibilities when undertaking management or restoration of coastal saltmarsh. As such, it introduces models for looking at the values of different saltmarsh states and the way in which, management and restoration bring about change.

# Chapter 4
# States and Values

## Describing Physical and Vegetative States

## 4.1 Introduction

Chapter 1 describes the processes associated with saltmarsh development. Chapter 2 provides a summary of the main human influences affecting change within the saltmarsh habitat. Chapter 3 sets out some of the changes in perception of the importance of 'natural' and semi-natural habitats to wider socio-economic values.

As pressures from natural forces and human actions change, the habitat responds (Adam 2002). This results in the saltmarsh occurring in a number of different 'states' depending on the pressures that are exerted on it. The condition of the environment (in this case the saltmarsh) provides the basis for identifying the 'state' of the habitat.

Each of these states has a different set of 'values'. These include a combination of ecological values, vegetation or animal interests, economic values, such as those associated with sea defence capability, and social values, including recreation and landscape. 'Value' in this context does not imply a fixed, quantifiable or objective measurement. Economic valuation (i.e. providing a monetary valuation of natural resources) is a specialist field, not dealt with here. Information that is more general is provided on the subject by the Organisation for Economic Cooperation and Development (OECD), see the Earthscan publications (e.g. Barde & Pearce 1991; Pearce et al. 1989; Pearce & Barbier 2000) and National Oceanic and Atmospheric Administration (NOAA) (Lipton et al. 1995). The 'values' assigned to each of the states are determined by its relationship to the conservation and management of the habitat. This chapter looks at the interactions between the driving forces and pressures affecting the states and values of saltmarshes.

### 4.1.1 Driving Forces, Pressures, States, Impacts and Response (DPSIR)

The DPSIR model adopted by the European Environment Agency (an extension of the **P**ressure **S**tate **R**esponse model, developed by Organisation for Economic Cooperation and Development) provides a causal framework for describing the

## 4.2 Physical States – Description

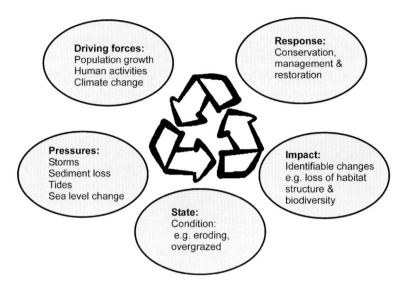

**Figure 25** The DPSIR model as it relates to saltmarsh. Note this only provides an illustrative list of factors influencing saltmarsh at each stage

interactions between society and the environment. **D**riving forces and resulting **P**ressures determine the **S**tate of the environment. Where changes occur which have an **I**mpact, there may be need for a policy or management **R**esponse (Figure 25).

Some of these values are associated with only one state, others may occur in more than one state. The states respond in different ways to the pressures exerted upon them and hence their values change. In order to help determine the most appropriate form of management it is important to know what the pressures are and how they influence the state and hence the value of the habitat. Taken together these help identify a response by way of management action, including habitat restoration, re-creation or creation.

In order to help unravel the complexities associated with conservation management and restoration of saltmarshes the discussion centres on two principal forms defined by their:

1. Physical state – referring to changes in the saltmarsh surface (vertical or horizontal) in relation to overall stability;
2. Vegetative state – closely related to the grazing management and transitions to terrestrial vegetation.

## 4.2 Physical States – Description

The presence of saltmarsh depends on a range of conditions described in Chapter 1. The establishment and growth of salt-tolerant plants on tidal flats is the precursor to saltmarsh development. The plants help bring stability to the surface as sediment accretion and the process of succession takes place. Thereafter, the nature of the saltmarsh is dependent on a variety of factors ranging from the availability of sediment, sea-level change and human interference.

In the absence of human interference, they grow vertically and horizontally until the factors leading to their establishment change (van de Koppel et al. 2005). Theoretically, saltmarshes will continue to grow naturally, through the deposition of sediment until only exceptionally high tides reach the upper levels. Transitions to other wetlands such as swamps and fens depend on drainage from the hinterland. Once away from tidal inundation, vegetation succession follows patterns similar to those of more terrestrial environments. Saltmarshes are restricted in their landward migration, where the land rises above the high water mark, either due to the local topography or because of the presence of sand dunes or shingle structures. Increases in sediment supply, decrease in wave action and storm activity, or lowering sea level will all tend to promote accretion rather than erosion. The reversal of those factors, facilitating vegetation establishment, limits the extent to which seaward growth is possible. Based on the interaction between these and other factors three main headings describe the status of the saltmarsh:

**State 1** Accretional
**State 2** Dynamic equilibrium
**State 3** Erosional

The two states (Accretional and Erosional) are not mutually exclusive. Erosion and accretion will occur on sites in a 'dynamic equilibrium'. It is important to recognise that describing a site as being in one state or the other, relates to the balance between accretion and erosion in the medium to long term. Short-term cycles of erosion, for example due to storms or changes in river channels, are part of the natural fluctuations influencing this dynamic habitat. Note that laterally eroding saltmarshes may also be accreting vertically. Although in the context of this analysis, they will normally be classified as 'erosional'.

## 4.2.1 State 1 – Accreting

Accreting saltmarshes are characterised by the establishment of plants on open tidal flats. Factors that bring this about include an abundance of sediment, sheltered locations where the fine particles can fall out of suspension, and salt-tolerant plants. The foreshore generally has a convex appearance, shelving gradually towards mean high water of neap tides, where saltmarsh formation usually begins.

The rate of seaward transgression and growth in height varies considerably from place to place. Recorded **horizontal** growth rates for the Wash, Lincolnshire are up to 50 m per year (Kestner 1962). At the same site a study of saltmarsh change from 1971–1985 showed average expansion rates of 20 m per year, although much higher rates occurred in front of recent enclosures (Hill & Randerson 1987). Estimates of lateral growth taken in Washington State indicated that clones of introduced *Spartina alterniflora*, expanded at approximately 0.5–1.7 m per year.

## 4.2 Physical States – Description

**Vertical** growth rates of 2–10 mm per year are normal for temperate saltmarshes in Europe and north-east America (Ranwell 1972). In a more recent study, rates of 3–11 mm per year were recorded from south-east England (French & Burningham 2003). Low intertidal saltmarshes in Washington and Oregon on the west coast of the USA had accretion rates ranging from 2.3–6.6 mm per year, with a mean of 3.6 mm per year (Thom 1992). Much higher rates of lateral expansion and vertical accretion occur when *Spartina anglica* invades mud and sand flats. Chapter 9 looks at the issue of *Spartina* in more detail. Accretion is a natural and essential part of the dynamic of the saltmarsh habitat (Figure 26).

The driving forces and the pressures affecting the saltmarsh (Table 5) represent natural events (1–3), effects resulting from activities occurring some distance away (4) and human activities acting directly on the saltmarsh (5–7).

**Figure 26** Expanding *Salicornia europaea* community, The Wash, Lincolnshire, UK. These communities wax and wane on an annual basis, though the general progression is for lateral expansion seawards, as biennial plants such as *Aster tripolium*, or perennials such as *Atriplex portulacoides* become established

**Table 5** Driving forces and pressures leading to saltmarsh accretion

| Driving forces | Pressures |
| --- | --- |
| 1. Balance between isostasy* and eustasy** | Sea level fall |
| 2. Sea cliff erosion | Increased sediment supply |
| 3. Increased precipitation and river flows | Increased sediment supply |
| 4. Deforestation | Increased sediment supply |
| 5. Managed realignment | Decreased tidal movement/wave action |
| 6. Sea defence e.g. warping, sediment fields | Increased sediment deposition |
| 7. *Spartina* invasion/planting | Increased rates colonisation |

*Isostasy, changes in land levels; **eustasy, global change in sea level

**Summary definition:** Accreting saltmarshes are those where the driving forces combine to encourage the progressive lateral expansion and vertical building of the habitat. Saltmarsh plants 'invading' tidal flats are an essential precursor to the development of saltmarsh habitat.

## 4.2.2 State 2 – Semi-Stable (Dynamic Equilibrium)

Mature saltmarshes will usually include both accretional and erosional phases of growth. When they are in a semi-stable or stable state, they can show patterns of erosion and accretion, which are largely in balance. As has already been intimated above, under natural conditions and with abundant sediment, accretion will usually dominate over erosion. However, as the saltmarsh matures it reaches a point of equilibrium. Thereafter, the forces driving erosion or accretion may cause cyclical change, but the overall area of saltmarsh remains more or less stable (in a state of dynamic equilibrium).

Saltmarshes respond to rising sea levels by accreting so long as the sedimentary and other environmental conditions are suitable. Vegetation aids sedimentation of tidal silts and creeks develop in response to tidal movement. Under the weight of sediment and other material compaction occurs giving room for further accretion. When relative sea level stabilises or falls, the amount of seaborne sediment decreases in proportion to internally generated organic material. At this point, with increased permeability and a decrease or absence of tidal flow, surface drainage features disappear. A reversal of this situation occurs if sea-level rise and or compaction continue to facilitate vertical growth. At any one site several stages in this development may occur, existing in a dynamic equilibrium with the environment, especially on larger saltmarshes (Allen 2000).

Other examples of cyclical change include those associated with tidal channels which deepen as accretion takes place. The deepening tidal creek results in increased tidal velocities and can cause erosion, releasing sediment and infilling the tidal creek again (Figure 27).

Movement of channels also influences the state of a saltmarsh. At one site on the River Kent, Morecambe Bay, in north-west England a saltmarsh eroded from 1,000 m wide to 150 m in 17 years (1975–1992). Here, the deeper water associated with a major channel, increased exposure of the saltmarsh edge to wave attack. As a result, channel shifts appeared to lead to cyclical change in this saltmarsh (Pringle 1995).

Change in the position of river channels is only one factor causing erosion on saltmarshes. Excluding anthropogenic activities (dealt with in Section 4.2.3) sea level change, storms and internally generated conditions, notably an increasingly steep shore profile can initiate erosion. In the early stages of development, there is a positive feedback between sediment accumulation and plant growth. With increasing maturity and height above the adjacent tidal flats, the profile becomes steeper and more vulnerable to wave attack. At this point disturbance from a storm or a change in channel position can 'tip' the saltmarsh into a phase of erosion. This may continue whether or not the initial forcing mechanism remains (van de Koppel

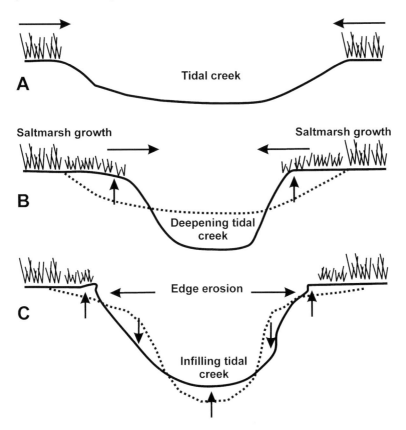

**Figure 27** A simplified diagram of the evolution of a tidal creek. A/B encroachment; C erosion, redrawn from Bird (1984). This may lead to the formation of cliffs and terraces when estuary channels meander from side to side

et al. 2005). The material derived from the eroding saltmarsh can accumulate and help re-establish the vegetation to form a series of steps (Figure 28).

Futher modification of the saltmarsh structure also takes place as other factors come into play, which change the pattern of the vegetated surface. Grazing is important not only affecting the type of vegetation, but also its physical strength (Chapter 7). Ice damage may cause a loss of vegetation in northern saltmarshes. Deposition of algae and other material on strandlines can smother vegetation, causing gaps to appear in the high marsh. Elevated salt levels in these areas may prevent recolonisation. Saltpans also develop and include two predominating forms:

1. 'Primary pans', originating early in the saltmarsh development as vegetation encloses small 'pools', which remain uncolonised by plants as the saltmarsh grows around them;
2. 'Channel pans' resulting from the enclosure of abandoned creeks (Pethick 1984).

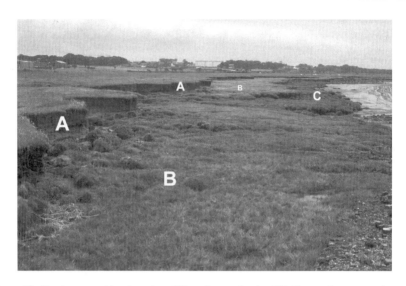

**Figure 28** Erosion caused by slumping cliffs at the marsh edge (**A**). Regrowth occurs and a series of steps develops as new saltmarsh extends seawards as each phase of erosion proceeds (**B**) and (**C**). An example from the Severn Estuary, UK

**Table 6** Driving forces and pressures, influencing the development of 'naturally' dynamic, semi-stable saltmarshes. Note, any one or more of these factors, if intense enough, can cause a permanent change in the state of the marsh, towards erosion or accretion

| Driving forces | Pressures |
| --- | --- |
| 1. Climate change | Increased storm activity |
| 2. Isostasy, eustasy | Sea-level change |
| 3. Natural geomorphological change | Cycles of change in winds and rain, ice cover, sediment delivery |
| 4. Changing populations natural herbivores (e.g. geese) | Loss in resilience as the vegetation is destroyed |
| 5. Death of saltmarsh plants, including algae | Smothering live vegetation |

Table 6 lists some of the factors, which help to influence natural change within a saltmarsh.

In some circumstances, these factors might lead to 'erosional' saltmarshes, particularly where sea level is rising against an elevated landward plain. Thus although the change is 'natural', because of the physical location of the saltmarsh there is no room to move and hence the saltmarsh state becomes 'erosional'. It is important to consider the timescales over which these changes occur. It may take several years before the balance of accretion over erosion becomes apparent. Making decisions to intervene based on one visit or one set of data is, therefore, not advisable.

Assessing this state can be difficult because of the timescales over, which the patterns change. For example, the development of saltpans may be more complex than described above, with timescales of tens of years required for their full development. Overall, the semi-stable state appears as a mosaic of saltmarsh

## 4.2 Physical States – Description

**Figure 29** Typical temperate mature saltmarsh showing saltpans with surrounding vegetation mosaics of mid to upper communities and a tidal creek in the background

vegetation, which includes a sequence of physical structures such as saltpans (Figure 29), dissected channels and saltmarsh creeks.

**Summary definition**: Saltmarshes are in dynamic equilibrium when sequences of erosion and accretion result in no overall change in their area, or the rise in saltmarsh elevation equates to a rise in sea level. This may involve periods when the saltmarsh front is moving landward or seaward depending on the influence of erosional versus accretional forces.

### 4.2.3 State 3 – Eroding

Erosion is a common state in many saltmarshes. Several factors can induce erosion. A reduction in sediment supply affects the resistance of the saltmarsh to waves, especially during storms. Sea-level rise increases the incidence and scale of waves. The enclosure of intertidal land and dredging contribute to the effects by reducing the tidal volume and increasing wave height respectively. Shipping use creates erosional waves. In Table 7 on the next page, the driving forces and pressures represent natural events (1–3), human activities acting directly on the saltmarsh (4–6) and effects resulting from activities occurring some distance away (7–9). They act on the saltmarsh singly or in combination and can turn an accreting, stable or semi-stable saltmarsh into one that is retreating.

The resulting erosion takes two main forms which affect the lateral extent of the saltmarsh, its height and surface topography:

- Lateral erosion

Lateral erosion occurs directly because of pressures exerted on the leading edge of the saltmarsh. Amongst these increased waves and/or tidal action in channels and creeks are significant factors. The example of channel movement (Figures 27 and 28) illustrates the nature of this type of erosion.

- Slumping

Lack of sediment and/or compaction can result in lowering of the saltmarsh surface, especially in areas where sea level is rising. Pollution (oil, litter and chemical), water logging and high levels of grazing and trampling cause internal rotting of the surface vegetation, further weakening the resistance to waves and tidal movement (Figure 30).

**Table 7** Driving forces and pressures leading to saltmarsh erosion

| Driving forces | Pressures |
| --- | --- |
| 1. Climate change | Increased storm activity |
| 2. Isostasy, eustasy | Sea level rise |
| 3. Rising populations of natural herbivores | Overgrazing (e.g. geese) |
| 4. Agricultural intensification, human activity | Increase in grazing/trampling |
| 5. Turf-cutting, borrow dikes and clay winning | Surface excavation |
| 6. Building earth banks, sea walls | Coastal squeeze |
| 7. Offshore sediment extraction, sea cliff protection | Reduced sediment supply |
| 8. Forestation, river damming | Reduced sediment supply |
| 9. Dredging/shipping use | Increased tidal movement/wave action |
| 10. Litter, oil pollution, eutrophication | Smothering vegetation |

**Figure 30** Erosion within the body of the saltmarsh can cause loss in height and result in surface changes. Factors that exacerbate the losses include snow-lie, ice rafting, water logging and smothering. The picture shows the loss of mature saltmarsh in the River Deben, Suffolk through a combination of lateral erosion and slumping

**Summary definition**: Saltmarsh classified as eroding, shows a pattern of change whereby the overall area of the habitat decreases with time. The nature of the erosion varies between sites and depends on the physical characteristics and more importantly the nature of the pressures effecting change. In this state, the saltmarsh diminishes in extent and tends to move progressively landward. As the saltmarsh area decreases so does its value.

## 4.3 Physical States – Values

Natural systems have important attributes that contribute to the welfare of human society. They also perform life-support services such as the purification of air and water, detoxification and decomposition of wastes, regulation of climate and regeneration of soil fertility (Daily et al. 1997). The values associated with these attributes, termed ecosystem values, include seafood, game animals, fodder and fuel, timber for building and pharmaceutical products. There is considerable literature on these concepts and their economic evaluation, see for example the web site http://www.ecosystemvaluation.org/index.html and the links and references included.

This book does not put a monetary value on saltmarsh, except in a few cases where information is readily available. It does attempt, however, to consider the contribution they make to the protection of life and property as well as cultural and nature conservation values.

Saltmarsh values begin to take shape as soon as the first plants take root in the tidal mud and sand flats. As the marsh grows, the values develop and become more complex. The states described above relate to the physical stability of the saltmarsh in relation to those factors promoting or arresting marsh development. At its simplest, the value of the habitat depends on its location and extent. A saltmarsh, which is accreting and forming a natural sea defence, will have a higher value than one that is eroding. Dynamic saltmarshes with a mosaic of physical structures and vegetation types are likely to have high biodiversity.

The description that follows attempts to identify the values of saltmarsh and the way these change as they move between the three principal physical states described in Sections 4.2.

### *4.3.1 Ecosystem Values*

Tidal saltmarshes rarely exist in isolation, forming an integral part of many estuaries, other tidal inlets and bays and deltas. To some extent, the values associated with them depend on location, size, and relationships with adjacent land and sea areas. Within these wider ecosystems, the principal values lie in the contribution they make to:

1. Natural processes and geomorphological patterns relating to highly specialised plants and animals found along the coastal fringe;
2. The trophic energy of the system through estuarine food chains and support for estuarine food webs (Figure 4);
3. The ability of the ecosystem recycling mechanisms for nutrients and other organic/inorganic compounds that in their turn contribute to the healthy functioning of the ecosystem.

These ecosystem values relate principally to the role that saltmarshes play in the functioning of the wider estuarine or other tidal embayment. They are well known and summarised in Chapter 1. Detailed descriptions of these ecosystem interactions are included in a variety of text books to which the reader is referred, see for example Adam (1990, Chapter 6); Packham & Willis (1997, Chapter 5.5), and in more general descriptions of the estuary environment, e.g. McLusky & Elliot (2004). Extrapolating these values in relation to economic, landscape or cultural interests is more difficult.

## *4.3.2 Economic Values*

Assigning an economic valuation to a saltmarsh is particularly difficult. In the context of this discussion the attributes are those which theoretically have a monetary value, e.g. as 'grazing land', or where the saltmarsh contributes to the economic well-being of the surrounding area. Historically saltmarsh enclosure has provided opportunities for development (Section 2.4 above). The new lands have provided clear economic benefits for agriculture, industrial and other infrastructure development. However, in the absence of enclosure, they also have other benefits that are often not readily recognised. The following sections attempt to provide some analysis of the type of benefits (values) that accrue to extant saltmarshes.

**Sea and flood defence** – Saltmarshes provide protection of the hinterland from waves and storms, through their contribution to sea defence. This is increasingly valued in the UK in relation to the maintenance of artificial sea defences (Brampton 1992; Toft et al. 1995).

This can have important financial implications in areas where the land is 'protected' by a sea wall or embankment, as it can greatly reduce the cost of installation and maintenance. Maintaining saltmarsh as an integral part of a defence can provide real financial benefit (Figure 31). Based on the estimates a 30 m wide saltmarsh could have a value, in monetary terms to the maintenance of existing sea walls of £6,000 per ha (King & Lester 1995).

Part of this value derives from the way saltmarshes attenuate waves. A study of wave attenuation over mud flats and saltmarshes in the Wash helped to confirm their value for sea defences. Using information gathered over a whole year the study quantified wave attenuation across intertidal surfaces (Coastal Geomorphological Partnership 2001). The results show that intertidal areas, with saltmarsh provide significantly greater wave attenuation than the adjacent mudflat (Table 8).

4.3 Physical States – Values

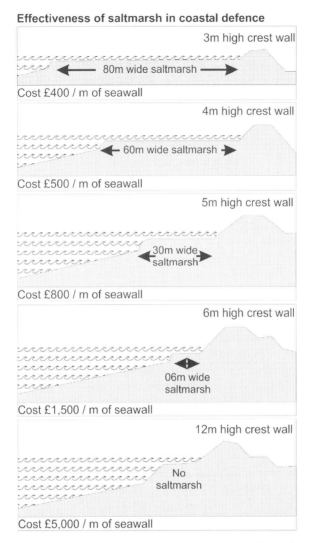

**Figure 31** Indicative costs of sea defences (early 1990s prices) with different lengths of fronting saltmarsh in south-east England (redrawn from Toft et al. 1995)

**Table 8** Wave attenuation across three saltmarsh sites in the Wash

| Transect | Dominant surface type | Average attenuation incident wave height (%) | Average attenuation incident wave energy (%) |
|---|---|---|---|
| Wrangle | Mudflat | 16 | 10 |
| Flats | Saltmarsh | 91 | 97 |
| Butterwick | Mudflat | 23 | 36 |
| Low | Saltmarsh | 64 | 72 |
| Breast Sand | Mudflat | 36 | 56 |
|  | Saltmarsh | 78 | 91 |

Further studies on saltmarshes in macrotidal areas, have confirmed and refined our understanding of the mechanisms involved. In North Norfolk, eastern England pioneer saltmarsh reduced wave energy by an average of 82% compared with 29% over a comparable width of tidal flat (Möller et al. 1999). On the Dengie peninsula, Essex also in eastern England, similar results showed 92% wave attenuation over a 310 m saltmarsh. Typically, the dissipation of wave energy occurred at an average rate of 0.5% per metre towards the land (more than twice that for unvegetated tidal flats). Most of this attenuation, >40% took place in the first 10 m, of permanently vegetated marsh (Möller et al. 2002; Möller & Spencer 2002). The passage of waves through saltmarsh vegetation is similarly reduced and wave height can be up to 71% and wave energy 92% lower in North American saltmarshes (Frey & Bason 1978). Based on this and other information the best intertidal profile for wave attenuation is one which has a relatively high topographic level and a wide, vegetated intertidal zone.

The type of vegetation may also be important. A saltmarsh plant with 'stiff' leaves dissipates wave energy, three times more effectively than plants with flexible leaves (Bouma et al. 2005). Data from one site on the Dengie peninsula also suggests a possible seasonal effect. The highest wave attenuation occurred in the autumn (maximum density of vegetation), with less attenuation in the winter months (when vegetation dies back) and least in the spring (before a full vegetation cover is re-established). This seasonality appears strongest in the pioneer zone, where colonising annuals fluctuate (Möller 2003). Saltmarshes with a high percentage of 'woody' perennial species, such as *Atriplex portulacoides*, appear more likely to provide greater resistance to waves. Higher-level saltmarsh, due to the shallower depth of water as the waves move landward and the increasing coverage of vegetation, may also increase the attenuation effect. Thus, the loss of the upper saltmarsh and transitional zones, through enclosure (Chapter 2), may be especially significant in reducing the resilience of the coast to erosion and flooding by the sea.

**Grazing** – The use of saltmarshes for grazing by domestic stock is a long established practice (Section 2.1.1.). It is difficult to put a monetary value on this management but they can support high densities of animals. In France, sheep grazed on tidal saltmarshes have a premium value. For example, in the Mount St Michelle, western France, up to 5,200 sheep graze the saltmarshes in the west, whilst cattle and horses graze the eastern part of the bay (Parlier et al. in press). Lamb from the saltmarshes of Normandy, is known as 'Pré-Salé' or Saltmarsh Lamb and has long been regarded as a delicacy in French restaurants.

In north-west England, where some of the most extensive and intensively grazed sites in Europe occur, stocking densities up to 6.5 sheep (year round) or 2 cows (summer) per ha have been recorded (Gray 1972). This lamb is sold as Pré-Salé lamb under the 'Holker Foods of Excellence' brand (source http://www.foodsofexcellence.co.uk/index2.htm).

Grazing also occurs extensively on 'natural' and constructed saltmarshes in the Wadden Sea and around the margins of those of the Baltic coast, helping to create short swards suitable for grazing ducks and geese (Dijkema 1984). Grazing levels equivalent to 9–10 sheep or 2–2.5 young cattle per ha (April – October) were recorded for other European marshes (Dijkema & Wolff 1983), which approach those of inland grassland. Whilst these levels are probably required for economic viability much lower levels (equivalent to 0.6 adult cattle per ha) are optimal for biodiversity (Kleyer et al. 2003). Grazing by domestic stock at least at moderate levels, has also taken place on many marshes in North America, although there is no equivalent information on stocking levels.

**Fish and shellfish** – The assertion that saltmarshes provide a major source of material for consumption outside the saltmarsh and are hence of direct economic value, particularly for commercial fish (as advocated by some of the early ecosystem studies) may be difficult to substantiate (Adam 1990, p. 348 ff). Most fish species may not feed directly, to any extent, on detritus produced by the saltmarsh plants or on food chains derived from such material. Despite this, they provide significant shelter from predation, which may in turn enhance the production of some commercial prey species (Boesch & Turner 1984). Whatever the precise relationship between saltmarsh primary productivity and fish and shellfish consumption, they play a vital role in some aspects of their life cycle.

A review of the use of intertidal saltmarsh by fish and macro crustaceans, found that both the biodiversity and biomass were significant. Around 15 species of fish and 6 species of crustacean used the Gulf of Mexico saltmarshes and 13 fish and crustacean used the Atlantic coast of south-eastern USA. Some of the species, which inhabited both the estuarine (saltmarsh) and marine environments, support important coastal fisheries (McIvor & Rozas 1996).

In the UK, juvenile Sea Bass *(Dicentrarchus labrax)* spend the first three or four years of their lives in estuaries, where saltmarsh forms an important component of their protection from predators. Protected nursery areas help to conserve the species, with fishing prohibited for all or part of the year within (mostly) estuarine areas. The value of this fishery and the importance of estuaries and estuarine saltmarsh to it, are reflected in the conservation measures supported by, for example, the UK Bass Anglers' Sport fishing Society (see http://ukbass.com/index.html).

The saltmarshes of the Wadden Sea are also of vital importance for the reproduction and life cycle of several fish species like the Atlantic Herring (*Clupea harengus*), European Plaice (*Pleuronectes platessa*) and Dover Sole (*Solea soleaI*). The mudflats and creeks associated with a saltmarsh in the Mira Estuary in Portugal, act as a nursery area for more than 40% of the fish species present in the estuary (Costa et al. 1988). These include detritus feeding species such as such as species of Grey Mullets (e.g. *Liza ramada*).

**Water quality improvement** – Studies on saltmarshes in the USA show that they can improve water quality. They act as a nitrogen interceptor, especially where they lie adjacent to highly permeable uplands. Here significant quantities of ground water reach coastal waters and hence reduce the amount of nitrate draining off the land. A simple diagram shows the pathways through which this operates (Figure 32).

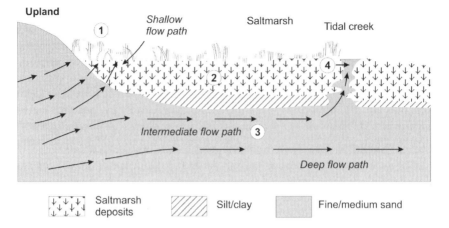

**Figure 32** Groundwater flow from the hinterland to saltmarsh. The study indicated four pathways for the movement of water: 1. boundary seepage zone; 2. interior saltmarsh zone; 3. subsaltmarsh aquifer; 4. creek bottom (redrawn from Howes et al. 1996)

As well as being sources and sinks of mineral nutrients and organic matter, saltmarshes also function as a sink for pollutants that would otherwise be damaging to the environment. Through these and other functions, they act as dynamic living filters for various, ecologically important materials. (Boorman 1999).

Other more localised values include:

- Providing **pipeline landfall sites** – the close proximity to the sea, ease of digging and relative remoteness makes them ideal for burying pipes. Restoration is also relatively easy;
- **Military training** – large remote open areas are often required for military training. Some major firing ranges lie over extensive saltmarshes (and tidal flats such as in the Wash in south-east England);
- **Samphire gathering** – *Salicornia* is a highly nutritious plant collected along the emerging marsh. It is an annual species that is collected and used in culinary dishes;
- Where transitions to upper marsh occur they also provide, **reed for thatching**;
- **Shooting** – they provide the basis for a hunting resource by providing food, shelter and 'safe' roosting areas for waterfowl.

### 4.3.3 *Cultural Values*

Saltmarshes have high scientific, landscape, recreational and cultural values. Although it is possible to assign monetary values to some of these, more often their worth is intangible. In this context the inspiration, they provide for literature, paintings or music has no monetary value even if the outputs have. Other uses have a more direct value, as in their use in education and research.

**Research and teaching** – Saltmarshes provide opportunities for studying 'natural' systems. Some of the earliest studies were on saltmarshes, which helped develop our understanding of ecology. Classic studies include Chapman working on the North Norfolk Coast; eastern England (e.g. Chapman 1938, 1939, 1941) reported in the Journal of Ecology. The description of the *Life and Death of a Saltmarsh* provides an enthralling account of the ecology and history of the habitat in the North America (Teal & Teal 1969). Ranwell's book on *Saltmarshes and Sand Dunes* (Ranwell 1972) is still a valuable textbook, helping provide a foundation for understanding how saltmarsh systems work.

**Recreation** – Recreational opportunities include bird watching, walking and nature trails, viewing areas and shooting. Saltmarshes also provide sites for boat mooring and other public access. In some instances, the habitat contributes to the overall experience of living. In the USA, for example, the 'Saltmarsh Alliance', a not-for-profit organization formed in 2002, helps support a Saltmarsh Nature Centre. Together with New York City's Parks and Recreation Department and the Urban Park Rangers, they supplement city funding enabling opportunities for nature trails and community-oriented events, which are valuable assets to the neighbourhood (see http://www.saltmarshalliance.org/).

**Landscape** – large saltmarshes can contribute significantly to the landscape in low-lying coastal areas, especially within estuaries. The open vistas often associated with these areas can enhance the tourism experience (Figure 33). They also provide a backdrop for paintings and a source of inspiration for poets and writers.

**Figure 33** Components of a 'day at the seaside', a car park, toilet, a beach for boating and paddling, set against a back drop of tidal saltmarsh and grazing marsh. Blakeney, North Norfolk

**Archaeological conservation** – Coastal lands provided favourable locations for early human settlement and saltmarshes provide archaeological evidence of this occupation. In south-east England, Pewit Island on the Essex coast contains the remains of an extensive oyster cultivation industry. Oysters were wintered in rectangular pits cut into the saltmarsh providing an important source of food for poverty-stricken Londoners in the mid nineteenth century (Strachan 1998). Today oyster fisheries remain on a smaller scale, although saltmarshes are not essential to their operation.

**Nature conservation** – Nature conservation values relate to their ecological status, including the presence of specialist halophytic plant communities and associated animals (especially invertebrates and breeding and wintering birds). It is important to recognise that this value changes depending on the status of the vegetation (Chapter 7). Some saltmarshes exist as nature reserves in their own right, but many more form part of wider protected coastal ecosystems. Within Ramsar sites saltmarshes 'Intertidal Marshes' were one of five most-frequently recorded wetland types in the regions of North America and Western Europe (Fraser 1999). Saltmarshes also form significant elements in many statutorily protected areas and nature reserves throughout the world. The influence of human activities (notably grazing) on the vegetation and the knock-on effects on other interests has a profound effect on their nature conservation value (Section 4.5).

## 4.4 Vegetative States – Description

Saltmarshes exist because of the establishment and growth of salt-tolerant plants. The factors affecting their physical state determine whether they grow, remain in a 'dynamic equilibrium' or recede (physical states 1–3, Section 4.2). One definition of saltmarsh is:

> "areas, vegetated by herbs, grasses or low shrubs, bordering saline water bodies." (Adam 1990).

This book deals with the vegetation of saltmarshes developing in tidal inlets and bays (estuaries etc. as described above and in Figure 2).

Human activities alter saltmarsh vegetation (Chapter 2). The effects range from total, irreversible destruction because of infrastructure development, to reversible enclosure for agriculture and grazing on unenclosed saltmarsh. Of these, grazing by domestic stock has a profound effect on the vegetation structure and species composition. Trampling can also cause localised damage to the physical structure through compaction and puddling. In turn, this has a significant influence on the nature conservation, as well as some other saltmarsh values. The state of the vegetation is described under four headings related to the number of domestic stock grazing on them. A fifth state is described, which results from overgrazing by native geese. This is principally associated with saltmarshes in North America and described in Section 7.4.

- State 1 – Heavily grazed;
- State 2 – Moderately grazed;
- State 3 – Ungrazed/Lightly grazed;
- State 4 – Abandoned, formerly grazed;
- State 5 – Overgrazed.

In the international Wadden Sea the intensity of grazing has been defined by the structure of vegetation (Bakker et al. 2005). The terms 'no grazing', 'moderate grazing', 'intensive grazing' and ' cutting' are defined by the canopy of the vegetation and its heterogeneity. They thus help describe the effective grazing situation in a specific area, irrespective of the stocking density, namely:

- Intensive grazing equal to an overall short sward;
- Moderate grazing equal to mosaics of low sward and tall canopy;
- No grazing equal to an overall tall canopy.

These states and the values attached to them, exist as a continuum. See Chapter 7 for a detailed discussion of the relationship between grazing levels, species composition and vegetation structure. Note that whilst State 1 and State 4 marshes are relatively distinct, the more moderately grazed examples (States 2 and 3) may merge into one another.

## 4.4.1 State 1 – Heavily Grazed

Heavily grazed saltmarshes are those, which have a history of grazing, usually by high numbers of domestic stock. The structural diversity of the marsh is reduced as most, if not all, of the standing crop is removed. At the same time, grazing eliminates sensitive species and favours tillering grasses. The latter include Common Saltmarsh-grass (*Puccinellia maritima*), which characterises the lower marsh in both non-grazed and in grazed areas, but extends into the mid-marsh in the more heavily sheep-grazed areas. For example in the Wadden Sea, 25 years of vegetation development recorded at permanent quadrats, showed that grazing negatively influenced *Atriplex portulacoides* and *Elytrigia atherica*, whereas *Puccinellia maritima* and *Festuca rubra* showed a positive response (Bos et al. 2002). In north-west England, high numbers of sheep produce a close-cropped sward dominated by the same species.

Saltmarsh classified as 'heavily grazed' usually supports grazing levels of up to 6.5 sheep (year round); 9–10 sheep or 2 cows (summer) per ha and represent some of the highest stocking rates recorded in north-west England and in the German Wadden Sea (Figure 34) respectively.

**Figure 34** Heavily sheep-grazed saltmarsh in the Dudden Estuary, Cumbria, north-west England. Note, the also complete absence of any structural diversity in the vegetation, which is a species-poor sward, dominated by *Puccinellia maritima*, *Armeria maritima*, Sea Plantain (*Plantago maritima*) and *Festuca rubra*

### 4.4.2 State 2–Moderately Grazed

Saltmarshes in north-west Europe moderately grazed by domestic stock, have a mosaic of vegetation with patches of tightly grazed swards (with *Festuca rubra* and other grasses) and denser more structurally diverse stands of other plants. These may include grazing-sensitive plants such as Sea Lavender (*Limonium* spp.) and/or *Atriplex portulacoides*. Stock densities will usually range from 5–6 sheep or 1.0–1.5 young cattle per ha. (April – October), approximately half those of heavily grazed saltmarshes. At these levels, there will be a reduction in the standing crop. At stock densities between this and 2.0 sheep or 0.3 cattle per ha (year round) they approach those for lightly grazed marshes.

Defining this state is difficult, since many variations occur in both time and space. Parts of some heavily grazed saltmarshes may have less heavily grazed areas, where access is restricted, e.g. by tidal creeks. Typically, the saltmarsh has a mosaic of short, medium and tall vegetation (Figure 35).

### 4.4.3 State 3 – Historically Ungrazed/Lightly Grazed

Historically ungrazed or lightly grazed saltmarshes are structurally diverse and tend to support higher biodiversity than other, more heavily grazed saltmarshes (State 1 or State 2). The greater diversity is reflected in the presence of grazing sensitive

## 4.4 Vegetative States – Description

**Figure 35** Moderately cattle-grazed saltmarsh, showing some structural diversity, with low-growing *Puccinellia maritima* and tussocks of high-level saltmarsh. Southern shore of the Solway Firth, Cumbria, England

species such as *Limonium vulgare*, which has a distinctive and attractive visible presence on many sites. Grazing levels do not usually exceed 2.0 sheep or 0.3 cattle year round. In northern latitudes where sheep grazing is common, such sites tend to be rare. They find their best expression on the high sandy shores of the south and east coasts of the UK, especially in North Norfolk, inaccessible areas of saltmarsh elsewhere and in the Mediterranean (Figure 36).

Note, historically ungrazed sites should not be confused or equated with sites where grazing has been abandoned (State 4). Here the changes in agricultural practice have led to a loss in overall biodiversity, see Section 4.4.4.

### 4.4.4 State 4 – Abandoned, Formerly Grazed

Saltmarshes grazed by domestic stock, and then 'abandoned' fall into this category. Reduction or elimination grazing pressure leads to the rapid growth of coarse grasses. This state is recorded from saltmarshes in England, Germany, the Netherlands and around the Baltic Sea where abandonment has occurred at many sites (Dijkema 1990). A detailed review on the Wadden island of Schiermonnikoog showed how Sea Couch (*Elytrigia atherica*) became dominant after 5–20 years on mature marshes,

**Figure 36** Ungrazed high-level lagoonal saltmarsh, Patokut Lagoon, Albania with abundant *Limonium* spp.

**Figure 37** Overgrown ungrazed saltmarsh (but formerly grazed) dominated by Sea Wormwood (*Seriphidium maritimum*), *Elytrigia atherica* and *Festuca rubra*. The Wash, Lincolnshire

slightly longer on younger marshes (van Wijnen et al. 1996). In the brackish marshes of the Baltic Sea reduction in cattle grazing led to the rapid spread of competitive species such as *Phragmites australis* and a decline in species richness and structural diversity of the saltmarsh (Lundberg 1996). The resulting vegetation tends to have a matted appearance with dense vegetation and low species diversity (Figure 37).

## 4.4.5 State 5 – Overgrazed

Overgrazed saltmarshes are those where grazing reaches a level that destroys the surface vegetation. This largely relates to areas (especially in the Arctic) where grazing by populations of native geese cause 'eat outs'. Although this has important consequences in those areas where it occurs, it is not a geographically widespread problem. Section 8.4.5 provides a more detailed consideration of restoring these areas.

## 4.5 Vegetation States – Values

Defining the values associated with the different vegetation states largely depends on their nature conservation interest. As grazing pressure has such a profound influence on these values, the following sections describe these in some detail. Note these values are not mutually exclusive and each may occur in more than one state, although they may have different relative importance. The way in which changes in grazing regimes affect other interests, is limited. For example, heavily grazed saltmarshes with little structural diversity, attenuate storm waves less well than those that are structurally diverse. The previous sections cover these values. The trends and trade-offs associated with the state of the vegetation in relation to grazing pressures is the focus of Chapter 7. What follows is a description of the values associated with the vegetative state of the saltmarsh.

### 4.5.1 Nature Conservation Values

Saltmarshes have a range of nature conservation values. These include the vegetation and the associated animals that live within the marsh for part or all of their life cycles, as described above. This section is concerned with the values linked specifically to grazing management. Generally, State-1 heavily grazed saltmarshes lack the plant species richness of the more moderately (State-2) or lightly grazed (State-3) sites, which tend to be biologically more diverse. What follows is a brief description of the range of biological interests associated with the habitat as a whole.

**Vegetation** – The plant communities present within an individual site provide one focus for the assessment and hence the selection of sites for nature conservation protection. Chosen examples reflect the main geographical types and completeness of the succession on an individual site. In Europe, for example, the geographical variation of the main vegetation communities provides the basis for the selection of Special Areas of Conservation (Table 9). Subdivisions within national boundaries modify the selection, reflecting local climatic and physical conditions. In Great Britain, there are 28 recognised community types and many more subcommunities (Rodwell 2000).

**Table 9** Coastal saltmarshes of Community interest forming the basis for the selection of Special Areas of Conservation (SAC's). The table includes the Natura 2000 codes and description taken from the Interpretation Manual of European Union Habitats (European Commission 2003)

| Codes | Directive Name | Description |
|---|---|---|
| Atlantic and continental saltmarshes and salt meadows | | |
| 1310 | *Salicornia* and other annuals colonising mud and sand | 'Formations composed mostly or predominantly of annuals, in particular Chenopodiaceae of the genus *Salicornia* or grasses, colonising periodically inundated muds and sands of marine or interior saltmarshes. *Thero-Salicornietea, Frankenietea pulverulentae, Saginetea maritimae*'. |
| 1320 | *Spartina* swards (*Spartinion maritimae*) | 'Perennial pioneer grasslands of coastal salt muds, formed by *Spartina* or similar grasses'. Including swards with native *Spartina maritima*, introduced *Spartina alterniflora* and the hybrid *Spartina anglica* |
| 1330 | Atlantic salt meadows (*Glauco-puccinelietalia maritimae*) | 'Salt meadows of Baltic, North Sea, English Channel and Atlantic shores'. A complex habitat with no less than 12 communities recognised in Great Britain (Rodwell 2000) |
| Mediterranean and thermo-Atlantic saltmarshes and salt meadow | | |
| 1410 | Mediterranean salt meadows (*Juncetalia maritimi*) | 'Various Mediterranean communities of the *Juncetalia maritimi*'. |
| 1420 | Mediterranean and scrubs thermo-Atlantic halophilous (*Sarcocornetea fruticosi*) | 'Perennial vegetation of marine saline muds (schorre) mainly composed of scrub, essentially with a Mediterranean-Atlantic distribution (*Salicornia, Limonium vulgare, Suaeda* and *Atriplex* communities) and belonging to the *Sarcocornetea fruticosi* class'. |

The specialist halophytic flora is obviously important, as are the rarer species with distinct geographical distributions. Amongst those in Europe are *Blymus rufus*, a species of northern latitudes and Shrubby Sea-blite (*Suaeda vera*) and several species of Sea Lavender such as Matted Sea-lavender (*Limonium bellidifolium*) with a much more southerly distribution, including the Mediterranean. For a more detailed description of the vegetation in temperate regions, see (Packham & Willis 1997, Chapter 5; Doody 2001, Section 5.3.1).

The saltmarshes of the eastern shore of North America differ from those of Western Europe in that two native *Spartina* spp. represent a major component of the vegetation, namely *Spartina alterniflora* and Saltmeadow Cord-grass (*S. patens*). Within the mid-Atlantic, coastal zone saltmarsh, *S. alterniflora* and Black Needlerush (*Juncus roemerianus*) are dominant. Spring tides irregularly inundate high marsh, which develops a savannah-like structure. The *Distichlis spicata* and *S. patens* dominate this high marsh zone. For a more detailed account of the worldwide distribution of saltmarsh, see Adam (1990, Chapter 3). These species support a wide range of animals, including breeding birds, see below.

## 4.5 Vegetation States – Values

As outlined above grazing has a significant effect on both the species composition and structure of the vegetation. This in turn significantly alters the nature conservation values on individual sites. A description of some of the more important features follows. These are summarised here to help set Chapter 7 on 'Trends and Trade-offs' between saltmarshes with differing grazing regimes, in context.

**Wintering waterfowl** – The short swards associated with heavily grazed saltmarsh are especially important for wintering ducks and geese, and large numbers of individual species can occur at any one site. In Europe wintering wildfowl include species such as Wigeon, Barnacle Goose (*Branta leucopsis*) and Dark-bellied Brent Goose, which feed directly on grassy saltmarshes and occur in thousands at some sites. Wintering Brent geese in the southern North Sea rely on saltmarsh for a significant proportion of their food requirements (Table 10). These species, and other herbivorous birds, rely particularly on heavily sheep-grazed saltmarsh swards. The geese, in particular, seek out open areas with close-cropped vegetation, not only for the palatable grasses, but also because they like to be able to see potential predators.

**Table 10** Three herbivorous wildfowl "Priority Species" and their habitat preferences; Saltmarsh Zone, **LS** – Lower Saltmarsh; **MS** – Mid Saltmarsh; **US** – Upper Saltmarsh (from Tucker & Evans 1997)

| Priority species | Zone | Plant cover | Vegetation height |
|---|---|---|---|
| Brent Goose (*Branta bernicla bernicla*) | US MS | Dense > 60% | Short – also feeds in the lower *Zostera* zone |
| Barnacle Goose (*Branta leucopsis*) | US MS | Dense > 60% | Short |
| Wigeon (*Anas penelope*) | US LS | Dense > 60% | Short – also feeds in the lower *Zostera* zone |

**Wintering passerines** – Saltmarshes in Europe also provide winter-feeding for a variety of small passerine species, amongst which the most numerous are Twite (*Carduelis flavirostris*) and Skylark (*Alauda arvensis*). They can occur in flocks of several 100s, along the coastal margins of the North Sea and the southern Baltic. These species tend to feed along the tide line or in the marsh, where accumulation of plant material, particularly seeds, occur. At a site in South Wales Skylark and Chaffinch (*Fringilla coelebs*) both feed on seeds with preference for communities with *Aster tripolium* and *Spartina anglica* (Kalejta-Summers 1997) species, which tend not to set seed when heavily grazed.

The Wadden Sea also supports important populations of wintering passerines such as Shorelark (*Eremophila alpestris*), Snow Bunting (*Plectrophenax nivalis*) and Twite. These species favour the lower saltmarsh vegetation and drift lines, although they also visit intensively sheep-grazed upper saltmarshes, which resemble lower saltmarshes in their plant composition.

**Breeding waterfowl** – The estimated breeding population of Redshank (*Tringa totanus*) on saltmarshes around the coast of Great Britain in 1985 and 1996 was 21,022 pairs and 16,433 pairs respectively. The two surveys were from 77 saltmarsh sites, and represented a substantial, approximately 45%, of the population breeding in Great Britain (Brindley et al. 1998). This species requires some structural diversity in the sward to provide camouflage for their nests and they tend, therefore, to nest more successfully in moderately or lightly grazed saltmarsh. Other ground-nesting birds, such as Oystercatcher (*Haematopus ostralegus*) and Lapwing (*Vanellus vanellus*) nest in more open situations. However, on grazed saltmarshes in northern Europe, they are susceptible to disturbance and trampling, which has important consequences for management.

**Breeding passerines** – A number of passerines breed in saltmarshes. In the USA, these include various species more or less restricted to the habitat. Several subspecies of 'Large-billed' Saltmarsh Savannah Sparrows (*Passerculus sandwichensis*) occur in coastal California and northwestern Mexico (Wheelwright & Rising 1993). Others species include:

- Coastal Plain Swamp Sparrow (*Melospiza geogiana nigrescens*) in north-eastern USA (Greenberg & Droege 1990);
- Three subspecies of Song Sparrows (*Melospiza melodia*) in tidal saltmarshes in San Francisco Bay, California (Arcese et al. 2002);
- The Eastern and California Black Rail (*Laterallus jamaicensis jamaicensis*) and (*L. j. coturniculus*) respectively (Eddleman et al. 1994; Conway et al. 2004);
- Breeding populations of the more common, Saltmarsh Sharp-tailed Sparrows (*Ammodramus cauducatus*), and Seaside Sparrows (*Ammodramus maritimus*), are two other specialist species of the saltmarsh of eastern USA;
- Two subspecies of the Marsh Wren (*Cistothorus palustris*) are of conservation concern in Florida (Greenberg et al. 2006).

In Europe, there are no species specific to saltmarsh, although several more wide-ranging species nest in high densities such as Skylark and Meadow Pipit (*Anthus pratensis*).

**Amphibians and reptiles** – In North America there are several subspecies of the Diamondback Terrapin (*Malaclemys terrapin*) restricted to saltmarshes in several states in the USA. The Gulf Saltmarsh Snake (*Nerodia fasciata* ssp. *Taeniata*) occurs on the Atlantic coast of Florida and two other snakes the Carolina Water Snake (*Nerodia sipedon*) and the Northern Brown Snake (*Storeria dekayi*) have subspecies living on saltmarshes (Greenberg et al. 2006).

**Invertebrates** – Various species of invertebrate inhabit saltmarsh, ranging from highly specialised species tolerant of rapid and prolonged tidal inundation, sometimes occurring in large numbers, to terrestrial species including spiders. In the absence of enclosure, saltmarshes with low levels of grazing by domestic stock have a good structural diversity, with several grazing-sensitive species such as *Atriplex portulacoides*, *Limonium vulgare* and Sea Wormwood (*Seriphidium maritimum*). These species are also important for some invertebrate animals which feed, or find shelter on the plants themselves. For more information, see Chapter 5, (Doody 2001 Section 5.3.4.) and (Adam 1990 Chapter 2, Invertebrate fauna).

A study in the Wadden Sea gives an indication of how the species composition changes along the saltmarsh gradient from pioneer to mature saltmarsh. Table 11 gives

## 4.5 Vegetation States – Values

**Table 11** Number of species of invertebrates found on different levels of saltmarsh in the Wadden Sea (after Dijkema 1984)

| Species | Mud/sand flat pioneer zone | Lower saltmarsh zone | Middle saltmarsh zone |
|---|---|---|---|
| Bottom species <1 mm | 300 | 350 | 1300 |
| Macro species | 100 | >500 | |
| Plant eating insects | 6 | 130 | 370 |
| Detritus feeding | – | 160 | 480 |
| Carnivorous | – | 80 | 200 |
| Parasitic | – | 80 | 250 |

a reasonable indication of the way in which invertebrate diversity increases towards the upper levels of the marsh. As might be expected there are few plant-eating species associated with the sand and mud flats and only three species found on each of the plants, *Spartina anglica* and Long-spiked Glasswort (*Salicornia dolichostachya*) but these increase with increasing plant diversity (Dijkema 1984).

The number of macrofauna is often dependent on the presence of specific host plants, changes in salinity and availability of detritus.

**Fish** – In addition to their value as spawning and nursery areas for some species of commercial fish (summarised in Section 4.3.2) they also support rare species such as the Saltmarsh Topminnow (*Fundulus jenkinsi*). This is one of several endangered species identified by the Florida Fish and Wildlife Conservation Commission (see http://myfwc.com/fishing/fishes/threatened.html), which is specifically associated with saltmarshes and brackish water.

**Mammals** – Saltmarshes support a few specialist mammal populations, as well more widespread and common species. In the USA, the Florida Saltmarsh Vole (*Microtus pennsylvanicus dukecampbelli*) is an extremely rare subspecies of the Northern Meadow Vole. As the name implies it is dependent on saltmarsh habitat, in this case dominated by *Distichlis spicata*. The Saltmarsh Harvest Mouse (*Reithrodontomys raviventris*) in San Francisco Bay is a California and federal endangered species, which is restricted to saltmarshes (Greenberg et al. 2006). Other more widespread species of voles can be common in rank growth of *Festuca rubra* in saltmarsh in the UK (Adam 1990).

# Chapter 5
# The Physical States

## Trends and Trade-Offs

## 5.1 Introduction

It is clear that saltmarshes have a range of values that change with the physical or biological conditions that influence them. The two general approaches to managing and restoring saltmarsh depend on whether the aim is to create or restore saltmarsh habitat, or manage existing saltmarsh vegetation. This Chapter deals with the first of these, which involves the trends and trade-offs associated with the creation, maintenance or restoration of the habitat. The principal concerns here relate to economic and social values, including those associated with flood protection (Section 4.3) as well as nature conservation. These include 'quantitative' parameters such as location, area, width and height of the saltmarsh. The 'qualitative' issues associated with nature conservation, such as the presence of specialist or rare plant and animal communities or vegetation complexity, are covered in Chapter 7.

Changes in those processes, which help promote saltmarsh accretion (sea-level change, tides and tidal range, sediment availability, freshwater flows and channel movements) also cause erosion. The two often exist in a dynamic interaction, with accreting (State 1) or eroding (State 3) occurring within the same site (Section 4.2.2). The balance between the two results in a landward or seaward progression of the saltmarsh front. Depending on this balance, management to promote the former and control the latter are key elements in any management or restoration strategy. This approach operates at a different scale to the manipulation of the vegetation (Chapter 7). It involves not only consideration of the saltmarsh itself, but also the influence of the wider estuarine environment.

This chapter looks at the nature of the processes causing erosion or accretion and the way in which the values associated with each state change, as the saltmarsh moves between them. The distinctions are not hard and fast or mutually exclusive, but represent convenient ways of evaluating appropriate management and restoration policies. This chapter provides a model based on these trends to help identify the most appropriate form of saltmarsh management or restoration. It is a complex process, which can involve promoting accretion through re-establishing surface stability and vegetation colonisation onto tidal flats. It may also include the protection and restoration of existing saltmarsh or alternatively re-creating saltmarsh on enclosed tidal land (managed realignment). Chapter 6 deals with the methods of management and restoration.

## 5.2 Physical Trends

Establishing the trend in the physical condition of the saltmarsh forms the basis for development of an evaluation model. In this model, erosion is the key factor used to distinguish between the states. It is the scale in relation to the saltmarsh as a whole that determines into which category the saltmarsh is placed. Whether a saltmarsh is eroding, stable or accreting also affects the contribution the saltmarsh makes to the estuarine ecosystem as a whole and hence, to many of the 'values' associated with it. In this context, the 'State Evaluation Model' also considers the value of saltmarsh to estuarine productivity and stability.

### 5.2.1 Processes Influencing the Physical State

External forcing factors (pressures) drive the saltmarsh processes towards one or other of the states. Hydrodynamic and sedimentary processes are important in determining the direction of movement. For example, in macrotidal saltmarshes on the French coast, lateral expansion was more prevalent when there was an abundance of new sediment. Relative sea-level rise or changes in hydrological conditions drove vertical accretion (Haslett et al. 2003). Saltmarsh vegetation also has an inherent resilience. However, soil conditions are important and the incidence of pollution may cause degradation of the vegetation surface. Grazing also affects the height of the sward and overgrazing can destroy the vegetation (Chapter 7). Understanding the way these interrelate, is important to the assessment of saltmarsh state as well as to any decision to alter the state. The more important of these interactions are summarised in Figures 38 and 39.

## 5.3 Values Associated with the Physical State of the Saltmarsh

Accreting saltmarsh (State 1) will have mostly positive values as new marsh ensures the continued existence of the habitat and with it its inherent values. Where erosion and accretion are in balance (State 2), within the saltmarsh there will be both structural and temporal change. These will include sites with significant cliff erosion (Section 4.2.2). These in turn will enhance some values, such as those associated with species diversity. The changes may be negative for some other values, such as sea defence, in some locations.

Overall, the value of the saltmarsh diminishes with time as erosion continues with no significant accretion (State 3). Thus most if not all of the attributes will

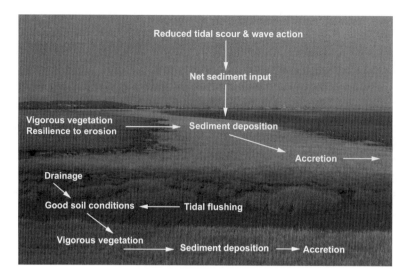

**Figure 38** Some of the key physical and hydrodynamic factors promoting vertical and lateral accretion within a saltmarsh

**Figure 39** Some of the key physical and hydrodynamic factors promoting erosion within a saltmarsh

## 5.3 Values Associated with the Physical State of the Saltmarsh

have negative values. Note that some of the positive values associated with accreting saltmarshes may also have negative impacts on other features, especially where lateral growth is rapid, as is the case with *Spartina* spp. expansion (Chapter 9). This section provides a summary of the values associated with each state and a description of the way they change with the changing state of the saltmarsh.

### 5.3.1 State 1–Accreting Saltmarsh

Where conditions are favourable, saltmarshes accrete. In the absence of human interference, this is the natural state of a saltmarsh, essential for the development of the full sequence of vegetation appropriate to the geographical region. Accreting saltmarsh is indicative of a healthy sediment budget and resilience to erosion. The largest saltmarshes tend to occur in meso- to macrotidal areas, with a net sediment budget and/or where relative sea levels are falling. They are associated with mostly positive or neutral trends (Table 12).

**Table 12** Positive or neutral trends in some of the values associated with accreting saltmarsh

| Value | Trend | Comment |
|---|---|---|
| Energy recycling mechanisms | Positive | Potential for greater productivity |
| Nutrient recycling mechanisms | Positive | Potential for greater efficiency |
| Water quality improvement | Positive | Potential for greater efficiency |
| Natural sea defence | Positive | Wider, greater resilience |
| Keep pace with sea-level rise | Positive | Maintaining or improving natural sea defence |
| Contribution to the landscape | Positive/Neutral | More expansive vistas |
| Ecological study | Positive | Presence of primary succession |
| Geomorphological study | Positive | Opportunities for sediment regime studies |
| Biodiversity | Positive/Neutral | Depending on the speed of succession and management. Note: Negative trends appear when the trend is towards dominance by a single species |
| Bird watching | Positive/Neutral | But see below – loss of mudflat |
| Walking | Positive/Neutral | |
| Samphire gathering | Positive | Though not where *Spartina* is the colonising species |
| Boat mooring | Neutral | Could be negative – more difficult access |
| Pipeline landfall sites | Neutral | |
| Military training areas | Neutral | |

**Figure 40** The foreshore along the quayside at Parkgate on the Dee Estuary in 1995. A former sandy beach, used for recreation in the 1920s

There are some negative aspects associated with accreting marshes. The rapid natural expansion of pioneer plants such as *Salicornia* spp., *Suaeda* spp. and in at least one case *Puccinellia maritima* (Edmonson et al. 2001) onto a sandy beach can have negative consequences for recreational activity. In the 1920s, children were able to paddle on open sandy intertidal sediments at Parkgate on the banks of the Dee Estuary (UK). By 1995, a saltmarsh some 1.2 km wide had developed in front of the promenade (Pye 1996; Figure 40). Siltation of the Ribble Estuary in north-west England (van der Wal et al. 2002) led to a similar problem on the Sefton coast, where a sandy beach became invaded by saltmarsh, resulting in *Spartina anglica* control (see Chapter 9). In the Wash, the Freiston foreshore hosted an annual 'summer Sand Fair' between the 1840s and 1870s (Robinson 1987). By 1980, the growth of saltmarsh was such as to allow enclosure and conversion to arable land. The loss of tidal flats can similarly be negative for wintering waterfowl, as new saltmarsh extends onto mudflats. This is particularly pertinent in the case of *Spartina* spp. when they occur outside their natural range or hybridise (Chapter 9).

## 5.3.2 State 2–Dynamically Stable

The state of 'dynamic equilibrium' is in many ways the desired state, especially in relation to nature conservation values. The functions of the mature saltmarsh from both a nature conservation and sea defence point of view appear to be mostly satisfied (Table 13). Thus, ecosystem values, economic values such as sea defence capability, use for grazing, contribution to fish and shellfish production or water quality improvement, all have positive values or at worse neutral values. This situation holds true so long as the overall status of the saltmarsh (area and elevation) remains in equilibrium. The temporal changes may also impart additional diversity. For example,

## 5.3 Values Associated with the Physical State of the Saltmarsh

**Table 13** Positive, neutral or negative trends in some of the values associated with dynamically stable saltmarsh

| Value | Trend | Comment |
|---|---|---|
| Energy recycling mechanisms | Positive/Neutral | Potential for increased rate of export to coastal waters |
| Nutrient recycling mechanisms | Neutral | ? |
| Water quality improvement | Neutral | ? |
| Natural sea defence | Positive/Neutral | More flexibility but areas of potential vulnerability |
| Keep pace with sea level rise | Positive/Neutral | More flexibility but areas of potential vulnerability |
| Contribution to the landscape | Positive/Neutral | Greater variation as features change |
| Ecological study | Positive | Presence of primary and secondary successions |
| Geomorphological study | Positive | Coastal processes in operation |
| Biodiversity | Positive | Greater number of species |
| Bird watching | Positive/Neutral | Availability of a wider range of feeding areas within the saltmarsh |
| Walking | Negative | Presence of wider and more frequent creeks |
| Boat mooring | Neutral | Could be negative – more difficult access |
| Samphire gathering | Neutral/Negative | |
| Pipeline landfall sites | Neutral | |
| Military training areas | Neutral | |
| Archaeology | Neutral | May expose and cover features of interest |

sequences of erosion followed by regrowth, create a series of steps as new saltmarsh develops to seaward of an eroding microcliff (Section 4.2.2; Figures 27 and 28). Each can have a different sequence of vegetation and hence biological diversity.

Saltpans add to the complex mosaic, and deposits of seaweed on the tide-line may smother the surface vegetation, creating further spatial variation as the strandline deposits rot (Packham & Willis 1997, pp. 101–105). The vertical structure of the vegetation provides further diversification, which helps to support a wider range of animals, especially invertebrates. In addition, the inherent dynamic nature of this state, also imparts an ability to respond to changing environmental circumstances, especially in relation to sea-level change.

### 5.3.3 State 3–Eroding Saltmarsh

The values associated with State 3, eroding saltmarsh, diminish as erosion takes place. As cliff erosion and in a few locations at the edge of the saltmarsh surface slumping, progressively reduce the saltmarsh area, most, if not all, of the attributes will have negative values (Table 14). Ultimately, all the values are lost, when the saltmarsh erosion reaches a point where only small remnants survive against a sea wall or rising ground.

**Table 14** Positive, neutral or negative trends in some of the values associated with eroding saltmarsh

| Value | Trend | Comment |
|---|---|---|
| Energy recycling mechanisms | Negative | Decreasing rate of export to coastal waters |
| Nutrient recycling mechanisms | Negative | |
| Water quality improvement | Negative | |
| Natural sea defence | Negative | Increase in vulnerability |
| Keep pace with sea-level rise | Negative | Lack of sediments |
| Contribution to the landscape | Neutral/Negative | |
| Ecological study | Negative | Presence of primary and secondary successions |
| Geomorphological study | Positive/Neutral | Coastal processes in operation |
| Biodiversity | Negative | |
| Bird watching | Positive/Negative | Wider tidal flats feeding areas, less breeding bird habitat |
| Walking | Negative | Presence of wider and more frequent creeks |
| Samphire gathering | Negative | No accreting saltmarsh |
| Boat mooring | Neutral/Positive | Could make access to tidal creeks easier |
| Pipeline landfall sites | Neutral/Negative | Reducing area for burial |
| Military training areas | Neutral | |
| Archaeology | Positive | May uncover features |

Despite the generally negative values associated with eroding saltmarsh there are at least two positive values. The first of these lies in the creation of larger areas of tidal sand or mud flats. These in turn increase the area available for intertidal invertebrates, prey for species of waterfowl and other predatory animals. Eroding saltmarsh can also expose former surfaces and features, which may have archaeological significance. These include submerged forests, in areas where sea levels were lower than today. In the context of this discussion, these may provide only limited compensation for the losses. Despite this, they represent a value worth considering when deciding whether to intervene or not.

## 5.4 Summary – A Physical Model for Change

It is possible to derive a State Evaluation Model taking each of the above states in turn. This seeks to summarise the key directions of change, and from this to analyse the relative merits of promoting saltmarsh accretion, protecting existing habitat or trying to reverse erosional trends. The approach involves assessing each state in relation to the principal concerns of the manager, whether for flood or sea defence,

## 5.4 Summary – A Physical Model for Change

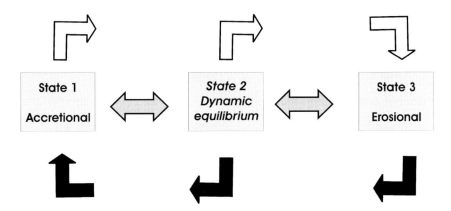

**Figure 41** A summary of the 'pathways for physical change' within an individual saltmarsh. Note 'creating erosion', i.e. moving directly from a State 1 to State 3, which mostly involves control of invading *Spartina* is not shown, but see Chapter 9

water quality objectives, cultural or socio-economic values and for nature conservation at any given site.

In order to determine the appropriate management option, it is important to assess the **desired** state in relation to the principal values affected by any move from the **existing** state of the saltmarsh. The 'Physical State Evaluation Model' provides a simple illustration of the possible directions of change in relation to any move from an accretional to an erosional saltmarsh (Figure 41). The model shows the main pathways for adverse change (open arrows) dealt with in this chapter and the 'routes' to restoration (black arrows) described in Chapter 6. Note that State 2 saltmarshes can incorporate both States 1 and 2, indicated by the grey arrows. Although these changes can lead to adverse effects, they are also part of the natural dynamic as described above.

The Physical State Evaluation Model identifies erosion as the principal reason for wishing to change the existing state of a saltmarsh. This involves reversing trends where erosional forces dominate over accretional ones. The trend is towards saltmarsh loss with accretion either non-existent or below that required to sustain the area of habitat in the medium to long term.

### 5.4.1 Rates of Accretion and Loss

If conditions are suitable, saltmarsh will accrete, even in areas where sea levels are rising. Vertical accretion rates in the outer Bay of Fundy, for example, ranged over the last two centuries from $1.3 \pm 0.4$ to $4.4 \pm 1.6$ mm per annum, similar to the recent rate of sea-level change recorded at Eastport, Maine (Chmura et al. 2001). In estuaries and other enclosed or partially enclosed embayments, where there is an abundance of sediment, rates of up to 7.7 and 6.3 mm per annum respectively have been recorded at the saltmarsh surface at Surville and Lessay on the West Cotentin Coast of Normandy. These are well in excess of the sea-level rise of 3.9 mm per annum determined from a tide gauge on the nearby island of Jersey, from 1952–2001 (Haslett et al. 2003). Vertical sedimentation over 5- and 40-year timescales in Blyth estuary, Suffolk, also easily outpaced a post-1964 sea-level rise of 2.4 mm per annum (French & Burningham 2003). In the Wadden Sea similar accretion rates of 5 mm and 10 mm per year were recorded over a 25-year period for barrier island and mainland saltmarshes respectively (Dijkema 1997)

Rates of expansion can be considerable under favourable conditions. The saltmarshes in the estuaries of the French coast are expanding laterally, in one estuary, at a rate of 4,400 m$^2$ per annum. This is attributed, partly at least, to the cessation of offshore sediment extraction and the presence of a more abundant supply (Haslett et al. 2003).

Despite the obvious ability of saltmarsh to accrete at a rate equal to, or greater than the rate of sea-level rise, there is little evidence for lateral accretion of new pioneer saltmarsh at many sites. In south-east England, there is a preponderance of erosion over accretion (Section 3.4.2). Saltmarshes in the Odiel Estuary, south-west Spain, show patterns of erosion exacerbated by human influences (Castillo et al. 2002).

Localised losses, in Jamaica Bay, New York, show erosion due to sea-level rise, accelerated by human activities, such as tidal dredging, which has increased the average water depth and with it tidal scour (Hartig & Gornitz 2001). Here major losses continue to occur because of increased waterlogging within saltmarsh interiors, slumping along their edges and widening of tidal inlets. Studies suggest that the losses will continue, as they appear likely to be unable to keep pace with accelerated rates of sea-level rise in the future (Hartig et al. 2002). Further evidence from the USA suggests that at its simplest, a saltmarsh that accretes at a rate greater than relative sea-level rise can keep pace with sea-level rise. One that cannot drowns (Schwimmer & Pizzuto 2000). The situation in the USA is particularly acute, as the combined effects of natural erosion and 'drowning' due to sea-level rise; exacerbate the more direct losses due to land enclosure and drainage.

Coastal wetlands are lumped together in the USA, as are the losses due to 'drowning' or human activity. For example in 1991, the Gulf of Mexico Coastal Wetlands included some 1,049,700 ha 'marshes' (fresh, brackish, and saltmarshes) (NOAA 1991). A review of a series of case studies provided information on the scale of losses within this area (Johnston et al. 1995). These included, Galveston Bay Marshes (fresh and non-fresh), which decreased from about 67,000 ha in the 1950s to about 52,800 ha in 1989, representing a net marsh loss of about 21% (White et al. 1993). For Coastal Louisiana, where coastal wetland loss represented 67% of the nation's total loss, some 177,625 ha were lost between 1978 and 1990. For the period 1956–1978, net wetland loss was even greater, at 267,800 ha. In Mobile Bay, non-freshwater marshes declined by 4,047 ha between 1955 and 1979; an overall loss of 35% (Roach et al. 1987).

There appear to be no worldwide estimates of saltmarsh loss due to erosion. The more localised studies in south-east England recorded direct losses through erosion (Section 3.4.2). Other studies, such as there are, refer to coastal erosion in general (e.g. in Europe the EURosion Project, see http://www.eurosion.org/). However, the losses highlighted above, coupled with the impact of direct human intervention (Chapters 2 and 3) and the effects of sea-level rise, which result in a saltmarsh squeeze (Section 3.5.2) help to confirm the view that eroding saltmarshes dominate over accreting ones.

## 5.5 Monitoring is an Essential Tool

Deciding on whether intervention is necessary, or desirable, leads to a first question for the manager: what is the current state of the saltmarsh? The description of the principal states (Section 4.2) provides the first level of assessment. It is possible in the short term to identify accreting or eroding saltmarshes. In most circumstances, it will be obvious. Cliffed edges appear in the saltmarshes, they may have slumping sides and where they occur, sea walls are undermined (Figure 42).

Determining the medium- to long-term trends requires monitoring. Anecdotal evidence may suffice when identifying potential problem areas, but it is more likely there will be a need for detailed work.

The position and area of a saltmarsh are the most obvious initial information requirements in any assessment. Remote-sensing using repeat aerial photographs,

**Figure 42** Eroding saltmarsh undermining a sea wall, Essex

and satellite images, or measurements from fixed-point markers, can provide first order assessments of change. The studies of change in the saltmarshes in Essex and North Kent using aerial photographs taken at different times (Burd 1992), helped determine a change in policy (Section 3.4.2). Satellite imagery has also proved useful in identifying long-term changes in the lateral extent of saltmarshes, in Jamaica Bay, New York (Wang & Christiano 2006). These involved identifying changes in relatively distinct, simple *Spartina* communities, other more complex communities are more challenging. In Australia, saltmarshes are important indicators for State of the Coast Reporting. Their extent can be mapped using aerial photography and satellite imagery. Ground-truthing helps differentiate between saltmarsh and areas of tidal mudflats (see http://www.ozestuaries.org/indicators/changes_saltmarsh_area.jsp).

Sequences of vegetation maps may be the only requirement when deciding the physical state of a saltmarsh. However, deciding on other forms of management (not just in relation to promoting accretion or controlling erosion), will require detailed survey and monitoring. In the Netherlands detailed 'Reference Conditions' have been established for saltmarshes in relation to the European Union Water Framework Directive. These include parameters for assessing changes in area and quality of the vegetation. The reference condition for the former is set against an estimate of historical acreages. The latter considers succession, zonation and quality characteristics. An assessment is made of the 'Potential Reference Condition' in relation to area and the 'Potential Good Ecological Status' of the vegetation (Dijkema et al. 2005).

The Massachusetts Office of Coastal Zone Management has produced 'A Volunteer's Handbook for Monitoring New England Saltmarshes' (see http://www.mass.gov/czm/volunteermarshmonitoring.htm). Aimed at helping local volunteer groups, this provides relatively simple approaches to collecting and recording data in a scientifically consistent way. The aim is to monitor the health of the saltmarsh as well as the effectiveness of protective measures and restoration actions. The Gulf of Maine Council on the Marine Environment has developed more rigorous standards (see http://www.gulfofmaine.org/habitatmonitoring/). The protocol for monitoring regional saltmarsh change, which also assesses the effectiveness of habitat restoration uses a tiered approach:

1. Tier I: minimal monitoring of core variables (hydrology, soils and sediments, and vegetation) occurring on most sites;
2. Tier II: recommended monitoring (Tier I plus one animal indicator (nekton, birds or invertebrates) where possible;
3. Tier III: intensive monitoring, all core variables, occurring at a small number of sites;
4. Tier IV: research into cause–effect relationships.

Whether this level of detail is required is a matter of judgement. It will inevitably depend on the available finance or other resources available in the area in relation to the 'value' of the assets at risk.

## 5.6 Assessing the Need for Intervention

The problem of assessing the balance of erosion over accretion is most difficult in larger sites. Erosion due to movement of tidal channels (Section 4.2.2) could suggest the need for intervention. Other factors such as seasonal patterns in wind direction, surface deposits, which smother vegetation or other natural forces that drive change, may cause short-term localised losses. Determining whether these represent short-term cycles or long-term trends may require both extensive and intensive survey. In the Tagus estuary, situated on the Atlantic coast of Portugal, despite losses due to human intervention the overall the balance of erosion over accretion was found to be neutral (Portela 2002). The loss trends identified in the USA due to the effects of 'drowning' caused by the consolidation of sediments, sinking due to tectonic movement, restriction of sediment supply or sea-level rise also appear clear. However, local considerations can result in the employment of inappropriate and counterproductive measures. It is therefore important that when making decisions, they take into account an estuary-wide perspective.

The need for intervention will also depend on the focus of the individual or organisation making the assessment. Whatever the actual change, intervention will depend on the social, economic or environmental reasons (singly or in combination) for taking action. The 'assets at risk' will in turn help determine the effort and financial commitment. As the nature of the saltmarsh changes so do the **values** attached to it, as described above. Thus changing the state requires an evaluation of the **desired state**, in relation to the **existing state**. The discussion that follows provides pointers to this evaluation.

### 5.6.1 Accreting – State 1

Accreting saltmarsh (State 1) is, under **most** circumstances desirable. It is positive, especially in relation to sea defence, or at worst neutral in respect of most of the attributes identified above and in Table 12. There is a **cost** as open tidal sand and mud flats are 'invaded'. Though even from an ornithological perspective, despite the potential loss of waterfowl feeding grounds it is unusual for a negative view to be taken, except in the case of *Spartina* invasion (Chapter 9).

**Benefits** also accrue in relation to the natural functioning of the ecosystem such as the contribution to primary productivity. Economic values (Section 4.3.2) and cultural assets (Section 4.3.3) are also favoured. In the absence of enclosure, it helps ensure the presence of all stages in vegetation development, through upper saltmarsh to brackish marsh and transitions to non-tidal vegetation. This in turn leads to the potential for the colonisation of animals at these higher levels, including breeding birds and invertebrates especially where grazing is moderate or light (Sections 4.4.2 and 4.4.3 respectively). In the absence of grazing succession can lead to the dominance of a single species, which is a particular issue in the Wadden Sea (Dijkema pers. comm.).

The expanding marsh will also provide greater opportunities for feeding and shelter for juvenile stages in fish development (as described in Section 4.3.2). From a management perspective, maintaining the conditions of sediment availability, protection from wave action to aid plant establishment and growth and preventing damaging developments such as enclosure, may be all that is required.

## 5.6.2 Dynamic Equilibrium – State 2

Saltmarshes in a state of 'dynamic equilibrium' will, under most circumstances also require no intervention. Here, the three States form a virtuous interaction with all stages represented within the one saltmarsh. Overall, they would normally show no overall change in total area in the medium to long term. Saltmarshes in a dynamic equilibrium hold a range of **values** (Table 13) associated with each of the principal 'States'. From a nature conservation perspective, they will have spatial and temporal variations that help to create a range of physical conditions suitable for a diverse flora and fauna. The landscape is likely to be similarly diverse. The values associated with economic interests, such as fisheries, are also catered for. Socio-economic, biological and nature conservation interests are all likely to be positive. From a coastal defence perspective, although small-scale erosion may cause problems locally, the dynamic nature of the habitat may be acceptable. Thus, from almost all perspectives, so long as the medium- to long-term assessment indicates no overall change, little or no intervention will be needed, under most circumstances. Note that where enclosure has taken place, even though the remaining saltmarsh may be in balance, issues such as sea defence may dictate intervention to re-create saltmarsh.

## 5.6.3 Eroding – State 3

Active intervention is more likely to be appropriate where saltmarshes are eroding on a wide geographical scale, or individual marshes are disappearing altogether. Eroding State 3 saltmarshes are undesirable under most circumstances, having negative values for most interests (Table 14). Reversing this trend (State 3 to State 1 saltmarsh) whether via State 2 or not, will be deemed to be positive (black arrows in Figure 41). This will represent a positive trend for those **values** associated with accreting or dynamically 'stable' saltmarsh.

Intervention to help encourage accretion is one course of action. Where this form of restoration is not possible, due to limitations on space or the suitability of the prevailing environmental conditions, there are other techniques. These involve protecting the remaining saltmarsh in situ, preventing further erosion by adopting protective measures. In other areas where erosion is severe, especially where sea level is rising relative to the land, saltmarsh re-creation may be the only option. Finally, if these fail, or are unsuitable then the creation of saltmarsh on land above the high water mark may be required. This may include terrestrial areas where the habitat has not occurred before. There are several approaches to reversing this trend.

## 5.7 Approaches to Restoration

Given the values recognised for saltmarsh and the positive trends associated with its restoration, it is not surprising that there are many examples of this activity. Chapter 6 describes the methods employed and their efficacy in some detail. At this point, it is sufficient to confirm that the restoration of saltmarsh is a clear benefit in securing sea defence, landscape and recreational, wildlife conservation and other objectives identified as having positive values and showing positive trends in Tables 12, 13 and 14.

In order to reverse erosional trends, which dominate in many areas and help to restore the values associated with the habitat there are several approaches:

- Moving seaward;
- Protecting and restoring saltmarsh;
- Moving landward, including creating new saltmarsh.

### 5.7.1 Moving Seaward, Creating New Saltmarsh

In the past, creating 'new saltmarsh' or intervening, to accelerate its accretion were common activities. According to the processes summarised in Section 1.3.1, given a reasonable supply of sediment new saltmarsh will become established or existing saltmarsh expand with or without human intervention. In the normal course of events, accreting saltmarshes (State 1) will eventually reach a mature state of dynamic equilibrium (State 2), when the forces promoting accretion are in balance with those promoting erosion.

The value of saltmarsh for sea and flood defence, even with relatively small widths of saltmarsh (Section 4.3.2) is clear. It may therefore appear that the creation of saltmarsh through the construction of sediment fields, 'warping' or *Spartina* planting (Section 2.4.3) will continue to represent a positive trend for coastal defence. The mechanisms designed to promote saltmarsh accretion onto tidal flats include reasonably well defined methods used in historical times and dealt with in Chapter 2.

The provision of biofuel, food and animal fodder are also reasons for saltmarsh creation. Scott et al. (1990) for example, reviews the use of *Spartina* as a biofuel. *Spartina* spp and *Phragmites* are both included as plants with the potential for providing biofuel material (Bassam 1998). Expanding the range of mangrove colonisation forms part of a practical approach to provide feed for sheep and goats and increase the Eritrean food supply (the Manzanar Project, see http://www.tamu.edu/ccbn/dewitt/manzanar/default.htm).

Promoting accretion in order to create **new saltmarsh** habitat, over bare tidal flats remains an option. However, there is a growing recognition that this is not an easy task, despite the inherent ability of saltmarshes to keep pace with sea-level rise. Human actions, notably saltmarsh enclosure, sediment deficits and the resulting foreshore steepening, can make establishment of pioneer plants difficult. In some cases, as in the case of planting non-native *Spartina* spp. it is undesirable (Chapter 9).

### 5.7.2 *Protecting and Restoring Saltmarsh*

Whilst the techniques for promoting accreting (State 1) saltmarshes are, in many cases, well-tried and tested, their sustainability, in all but the most favourable circumstances, is less clear-cut. Protecting existing saltmarsh represents a second option in any programme designed to reverse erosional trends. On the face on it, these appear to be wholly positive and desirable activities, especially when flood protection is a key issue. Preventing damage or loss of habitat, caused by human intervention also lies at the core of any nature conservation effort. However, a question arises about the sustainability of individual actions designed to prevent erosion, particularly in the face of rising sea levels, adverse hydrological conditions or depleted sediment supply.

The most commonly used techniques involve attempts to protect surviving habitat, repairing or restoring eroded saltmarsh vegetation. The methods include a variety of approaches including the erection of protective structures seaward of any remaining saltmarsh and replacing lost saltmarsh, 'in situ' through sediment placement and planting. The discussion in Chapter 3 shows how attempts to protect eroding saltmarsh used the 'warping' techniques borrowed from the Wadden Sea (Section 2.4.3) largely failed when applied to the eroding saltmarshes of the Essex coast (Section 3.5.1). Nevertheless, techniques to protect saltmarshes from erosion remain part of the armoury of measures employed throughout the world (Figure 43 and Section 6.2).

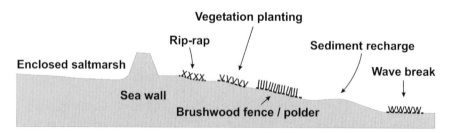

**Figure 43** Summary of some of the techniques for saltmarsh restoration seaward of an existing sea wall

Due to the extent of the losses resulting from human activities (Chapter 2), coupled with erosion forcing 'saltmarsh squeeze', simply maintaining the existing saltmarsh (what is left) may be insufficient. In order to retain the saltmarsh values, such as those associated with sea defence, requires re-creation of the habitat. Note the distinction between activities that seek to protect or restore saltmarsh where it has eroded from tidal flats (restoration, Figure 43) and those areas involving re-integration with enclosed tidal land, which is dealt with next.

### 5.7.3  *Moving Landward, Re-integration and Habitat Creation*

Many of the historical techniques for restoring saltmarsh are well known. Reversing the process of enclosure through re-integration with the sea involves a relatively new set of approaches (Figure 44). These include removal of all or part of an enclosing sea wall, embankment or dyke, originally built to create new land for agriculture or other development (Section 2.4). Note that the nature of the use of the land following the original enclosure is critical to the ability to re-create saltmarsh. Areas remaining below high water without buildings, roads and the like, such as land in agricultural use, have the greatest potential for re-integration. The precise methodology depends on the physical situation. However, some or all of the approaches summarised in Figure 44 are used. In some circumstances, built structures such as roads need not be an impediment to re-integration. In the USA, for example, enlarging culverts under roads and removing tidal flaps may be all that is needed.

Decisions to create, protect or restore saltmarsh will need to have information on the status of the sediment regime, tidal range, exposure, relative sea-level change (rising or falling), efficacy of the techniques and the 'knock-on' effects for other interests. Having made a decision to restore saltmarsh, there are several methods. Chapter 6 considers these as they apply to restoring saltmarsh (re-creation and re-integration) and creating new saltmarsh.

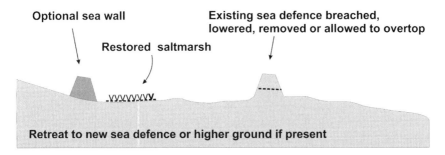

**Figure 44** Summary of some of the approaches to the landward restoration of saltmarshes and tidal flats

# Chapter 6
# Physical States, Restoration Methods

## Re-creation, Re-integration or Creation

## 6.1 Introduction

Chapter 5 provides information on the main reasons why an individual manager, compliance operator or organisation might wish to change the physical state of a saltmarsh. For the purpose of this discussion, there are several different approaches:

1. Promoting accretion;
2. Intervening to protect eroding saltmarsh;
3. Restoring eroding habitat;
4. Allowing nature to take its course (leaving things alone);
5. Re-creating lost habitat;
6. Creating new habitat (where non existed before);
7. Promoting erosion (loss of saltmarsh), mainly concerning non-native *Spartina* spp. dealt with in Chapter 9.

This chapter provides a summary of the main methods of restoration, i.e. the established ways of moving from the 'existing state' where saltmarsh is lost, to the 'desired state' where saltmarsh is restored. Two main headings reflect the approaches described in Chapter 5 (Figures 42 and 43), namely:

1. Restoration within the existing (unenclosed) tidal frame, mostly involving 1, 2 and 3 above;
2. Re-creation of (enclosed) tidal land, mostly involving 4, 5 and 6 above.

## 6.2 Restoring Eroding (State 3) Saltmarsh, Moving Seaward or 'Staying Put'

Re-creating saltmarsh habitat methods considered in this section are concerned largely with encouraging accretion of sediments and plant establishment on open tidal flats. There is a vast literature relating to this especially from the USA. The following sections introduce the methods employed to promote vegetation development. These include:

- Warping;
- Bay bottom terracing (in the USA);
- Use of dredged material;
- Reseeding and other vegetation restoration;
- Planting Cord-grass (*Spartina* spp.).

These methods may be accompanied by protective structures to reduce wave attack such as:

- Offshore breakwaters;
- Use of 'rip-rap'.

## 6.2.1 Warping, Poldering

These methods include those derived from the historical approaches to saltmarsh enclosure for use as agricultural land, as described in Chapter 2. The extensive warping adopted in the southern North Sea, especially in the Wadden Sea, the 'Schleswick–Holstein' method (Section 2.4.3; Kamps 1962) and in south-east England (Carey & Oliver 1918) created new, unenclosed saltmarsh. This process is labour-intensive and all but ceased in most areas of the Wadden Sea after the 1940s (Beeftink 1977a). The creation and maintenance of these areas were important for sea defence and continued up to 1992, in Denmark at least (Figure 45). Grazing follows as the saltmarsh vegetation develops, to create short, species-poor swards (Figure 46).

**Figure 45** Parallel lines of excavated ditches enclosed by brushwood fences, designed to encourage saltmarsh development seaward of a sea wall in the Danish Wadden Sea, 1992. Note excavated ridges to encourage vegetation

**Figure 46** 'Mature' grazed saltmarsh derived from 'warping' at Beltringharder, German Wadden Sea, 1988

The sediment fields provide protection from erosion and work best where there is a fetch of <200 m. Trenches create raised areas encouraging saltmarsh plants to become established. In the UK, experimental attempts to protect eroding saltmarshes, used these techniques borrowed from the Wadden Sea. In Essex, southeast England these approaches were largely ineffectual (Section 3.5.1). Some of the original polder-type structures are still in place but show little or no signs of developing saltmarsh (Figure 23). The experiments are in abeyance.

### 6.2.2 Bay Bottom Terracing in the USA

This is a process used in North America, especially around the Gulf of Mexico. The method also borrows the approach adopted in the 'Schleswick–Holstein' method. Instead of creating 'sediment fields' using brushwood groins, 'terraces' are built using material dredged from the bay bottom. Planting with *Spartina alterniflora* takes place on the terraces. The Sabine Terracing Project (Baton Rouge, Louisiana) showed how these become colonised, with the vegetation spreading laterally from the raised terraces. Although limited sediment accumulation took place within the 'cells', they did not become vegetated. Other similar projects required additional structures, including geotextile tubes. Discussion of the merits of this approach, suggests that restoration using the terracing method may be useful in shallow water in the northern Gulf of Mexico. However, the technique does not seem suited to the creation of large expanses of saltmarsh (Turner & Streever 2002).

### 6.2.3 Use of Dredged Material

Dredging is commonplace in estuaries and coastal embayments, where navigation channels are required for the passage of ships. Prior to 1970, in the USA the primary use of dredged materials was to build or expand land for airports, ports, residential, or commercial development. Where such uses were not established, the material was often taken offshore to be dumped and lost to the system altogether. As environmental issues became more important, disposing of dredged material 'in a beneficial way' resulted from the need to give greater consideration of its environmental effects (US Corps of Engineers 1987). More specifically, the value of shorelines for flood defence and coastal protection helped promote its use in coastal management.

The US Environmental Protection Agency and the US Army Corps of Engineers host a web site that describes the general approaches when using dredged material, for 'agricultural/product uses, engineered uses and environmental enhancement' (see, http://el.erdc.usace.army.mil/dots/budm/budm.cfm).

The methods employed in this form of restoration vary considerably. They depend on the nature of the original material, including its quality and quantity, as well as the distance to, and exposure of the disposal site. Hydraulic-pumping, designed to increase the elevations of the intertidal to mean sea level, is the usual form of operation. In open, exposed sites, this may be accompanied by protection using rip-rap or breakwaters. After the deposits de-water, vegetation planting can take place, although at many sites natural colonisation can be successful. In the Sabine National Wildlife Refuge, for example, four areas became colonised initially by *Spartina alterniflora* and after 5–7 years by species more typical of higher levels, including *S. patens* (Turner & Streever 2002).

A question arises as to the effectiveness of these approaches, including the extent to which the restored or created saltmarshes match 'natural' ones. Vegetation structural characteristics may take only a few years to become similar to those in the natural marshes. However, although older created saltmarshes in south-west Louisiana had greater organic matter than younger ones, even after 19 years, this was significantly lower than adjacent natural saltmarshes. Whilst vegetation structure takes only a few years to match 'natural' saltmarshes, 'it takes several decades for the soil characteristics to reach equivalency with the natural marshes, if they ever will' (Edwards & Proffitt 2003). Saltmarshes develop through a complex range of physical and biological interactions (Bakker et al. 1997); it is perhaps not surprising that it takes time for the processes to re-create soil characteristics and plant communities equivalent to unmodified areas.

Other studies of the productivity of wetlands, suggest that it may take up to 35 years for net annual aboveground primary productivity to reach that of natural saltmarshes (Edwards & Mills 2005). Key factors in using dredged material to create or re-create saltmarshes lie in the elevation and structural diversity, which should be as similar as possible to natural saltmarshes (Edwards & Mills 2005). A review of the literature suggests that dredged material marshes 'provide some of the functions of natural marshes, but probably do not replace all of the functions of lost natural marshes.' (Streever 2000).

### 6.2.4 Reseeding and other Vegetation Restoration

The establishment of saltmarsh vegetation depends on the availability of sediment together with the ability of salt-tolerant plants to become rooted in the intertidal zone (Section 1.3). Similarly, the introduction of seeds or vegetative shoots or seedlings, require conditions which facilitate plant establishment. The desirable conditions are:

- Protection from wind and waves;
- In positions of restricted tidal scour;
- Located between mean sea level and high water;
- A slope of between 3–5%;
- Sediment rates of between 3–10mm per year;
- Firm, oxygenated silt, the smaller the grain size the more suitable for lower elevations.

Natural colonisation will be the best approach, since it is more likely to provide a 'match' when attempting to re-create 'natural' looking saltmarshes. Planting may be appropriate in some circumstances, but requires good knowledge of the type of vegetation derived from descriptions of nearby saltmarsh. Information on the life requirements and suitability for planting naturally occurring saltmarsh plants provides the basis for vegetation restoration. In the UK, at least 22 species are potentially suitable for flood-defence engineering works. These cover the principal communities, representative of their position in the tidal frame (Brooke et al. 1999).

### 6.2.5 Planting Cord Grass **Spartina** *spp. in the USA*

Extensive planting of *Spartina* spp. was one of the earliest forms of saltmarsh restoration in the USA, dating back to the 1950s in Chesapeake Bay, Virginia (Broome 1990). It was, and often is, used in combination with some of the physical techniques described above. Studies in Galveston Bay suggested protection from waves was necessary before the establishment of transplants below the normal high tide could be achieved (Webb & Dodd 1983). Native species of grasses (primarily *Spartina alterniflora* and *S. patens*) are commonly used. [Note that this section is concerned with re-creating saltmarsh, using plants native to the area for the purposes of reversing habitat loss. Planting non-native *Spartina*, formerly practised as part of a process involving enclosure and land claim, is inappropriate (see Chapter 9)].

*Spartina alterniflora* successfully established on terraces created in the Sabine Terracing Project, northern Gulf of Mexico. Both clumps and individual plants colonised rapidly and there were no apparent differences in survivorship and growth after two years (Turner & Streever 2002). Experiments in spacing of *Spartina alterniflora* plants along an eroding North Carolina shoreline over a ten-year period showed that the transplanted saltmarsh was equal in primary productivity that was 'persistent and self sustaining'. On marginal sites near the lower tidal tolerance limits, 45 cm and 60 cm spacing was needed, whilst 90 cm spacing was adequate under favourable growing conditions (Broome et al. 1986). Other factors important for successful transplantation include the time of planting. Planting *S. alterniflora* on sandy dredged material on the Bolivar Peninsula, Galveston Bay, Texas, was more successful in May than February, when higher tides shifted the zone of survival upwards (Webb & Dodd 1989).

Saltmarsh restoration is often associated with attempts to involve the local population, especially in the USA. 'Restore America's Wetlands' is a wide-ranging programme, which seeks to support the 'Estuary Restoration Act of 2000' through a 'Community-based Restoration Programme' (http://www.estuaries.org/). There are many other initiatives, some supported by the US Environmental Protection Agency, such as the Gulf of Mexico Programme (see http://www.epa.gov/gmpo/welcome.html). Although these cover wetlands in general, individual projects include saltmarsh restoration.

The situation in Narragansett Bay, Rhode Island, provides an illustration. Here one of the factors affecting the future of the Narragansett Bay environment is the condition of its saltmarshes. A 'Save the Bay' initiative, works with communities and community groups to protect and restore saltmarshes, constantly threatened by poor resource management and neglect. Bay staff work with local volunteers to plant a variety of saltmarsh plants (including native *Spartina*) at a number of restoration sites, (see http://www.savebay.org/habitat_saltmarshrestoration.asp). The intensity and scale of the restoration process can be immense. For example, in Tampa Bay, Florida, 250–300 volunteers from a wide variety of groups set about planting between 20,000–30,000 plugs of saltmarsh grasses in one day (see http://www.tampabaywatch.org/index.cfm?fuseaction=content.home&pageID=23).

### 6.2.6 Offshore Breakwaters

Although saltmarsh will establish naturally, this only occurs if the water is sufficiently quiescent to allow the plants to become rooted in the tidal mud and sand flats. Protecting exposed shores with a variety of offshore structures provides one means of helping plants to become established. In the Gulf of Mexico, plantings behind offshore breakwaters were mostly successful, whilst exposed (unprotected) shores were not. A variety of 'soft' engineering approaches such as the use of sand bags, were cost effective, when compared with more conventional 'hard' engineered structures (Allen et al. 1990).

**Figure 47** Construction of shore-linking groynes 1989, Sales Point, Dengie, Essex, UK. In the distance, Thames barges sunken in situ

In the UK, Thames barges, filled with sand in situ formed part of an attempt to prevent erosion (estimated at 10%, over a 21-year period; Harmsworth & Long 1986) and aid the colonisation of 'sediment fields' at three sites. At each site, on the Dengie peninsula, Essex, UK, the approach was slightly different:

1. **Sales Point** – In 1986, 11 Thames barges placed 200 m offshore, were spaced 20 m apart, to create a wave break 'protecting' 600 m of eroding saltmarsh. In 1989 brushwood groynes were erected, connecting the lighters to the shore (Figure 47);
2. **Marsh House** – In 1984, 16 Thames lighters placed 500 m offshore, were spaced 20 m apart. The construction of two groynes (clad with geotextile material rather than the traditional brushwood) in 1986 provided protection at either end of the wave break. *Spartina* was planted in 'gripped' areas towards the landward side of the polder;
3. **Deal Hall** – 2,400 m square polders, constructed in 1980, used brushwood groynes. The enclosed tidal mudflats were 'gripped' in 1981 and again in 1989. A third polder, constructed in 1987/88 was not 'gripped'.

Preliminary reviews and descriptions of these, and other experimental sites in Essex, suggested that their effectiveness was patchy (Holder & Burd 1990). In 2003, the saltmarsh section of the Dengie Site of Special Scientific Interest was eroding and in an 'unfavourable' condition. "Detailed saltmarsh survey data shows significant erosion of saltmarsh" (Site Condition Report, available from within the Natural England web site (see http://www.english-nature.org.uk).

**Figure 48** 'Rip-rap' – introducing hard material to protect eroding saltmarsh; intrusive feature or successful management? Severn Estuary, Gwent, South Wales

## 6.2.7 Rip-Rap, Protecting the Eroding Edge

Material placed directly on the eroding edge of a saltmarsh also provides another option for protection (Figure 48). This detracts from the appearance and introduces a wholly unnatural element into the system. This probably only appeared to work because of the cyclical nature of the lateral expansion of the saltmarsh. This expansion being a response to changes in the erosion patterns of the saltmarsh (Figure 28).

## 6.2.8 Setbacks to Planting Native Spartina

Saltmarsh restoration can be an extensive and successful activity, whether in association with physical structures or not. However, re-creating saltmarsh vegetation is not always straightforward. In Coastal Louisiana, planting *Spartina alterniflora* to overcome losses resulting from enclosure and agricultural use, suffered 'die back'. The losses, called 'brown marsh phenomenon', primarily involved the rapid browning and 'die back' of *Spartina alterniflora*. This prompted a special conference to look at the problem, which was particularly prevalent in the year 2000. A variety of causes were thought to be responsible including marginally high salinity levels; possibly resulting from drought, unusually high summer temperatures, reduced rainfall and lowered water levels (Stewart et al. 2001).

A similar situation affected *Spartina alterniflora* and Needlegrass Rush (*Juncus roemerianus*) saltmarshes in Georgia. In 2001 and 2002, more than 800 ha throughout the coastal zone exhibited 'die back'. Results from a study of plants transplanted, from May to October 2003 into 'die back areas' suggested, this might be a transient phenomenon. Healthy plants survived and transplants continued to be an option for restoring affected areas (Ogburn & Alber 2006).

Planting *Spartina patens* in Tampa Bay, Florida seemed to have low survivorship due to 'transplant shock'. An investigation into this suggested that a combination of low redox potentials and elevation were significant factors. As plant health varied more with elevation than redox potential, avoiding planting in highly reduced areas, and increasing planting to a higher level, would increase survivorship (Anastasiou & Brooks 2003).

## 6.3 Restoring Saltmarsh – Moving Landward: 'Re-integration with the Sea'

The previous sections describe methods seeking to encourage the growth of saltmarsh through the restoration of eroding vegetation, or even to establish new habitat on tidal flats. The alternative considered here is to move landward, re-creating habitat on tidally restricted or enclosed former tidal land, converted to agricultural use. The former was prevalent in the USA and adopted for southern New England saltmarshes from the late 1970s. The latter is a relatively new and innovative approach in the UK, but also adopted elsewhere in Europe. It involves giving up land to the sea, much of which is in agricultural use following enclosure (Section 2.4.2). This section looks at these approaches, which recognise that restoring or creating saltmarsh on remaining tidal flats may be unsustainable in the medium to long term, especially in areas where sea level is rising relative to the land.

### *6.3.1 Restoring Tidally Restricted Saltmarshes in the USA*

Restricted flows using tide gates, associated causeways and dikes provided flood protection, mosquito control, and/or land for salt-hay farming for several decades, in the USA. This caused deterioration in the quality of the saltmarsh meadows with *Phragmites australis* and species of Bulrush (Cattail) *Typha angustifolia* and *T. latifolia* replacing the native *Spartina* spp. Studies in Connecticut suggested that large-scale restoration, through the reintroduction of tidal flows, would help re-establish the role of saltmarshes in the estuarine ecosystem (Roman et al. 1984). Over a period of 20 years, some of the successful methods for habitat re-creation in Long Island Sound, Connecticut, centred on removal or enlargement of tidal culverts, abandonment or modifications to tide gates, excavation of tidal channels and other water control structures, to re-create historic tidal flushing regimes (Warren et al. 2002).

Other successful schemes, such as Sachuest saltmarsh within Narragansett Bay, involved the replacement of a narrow pipe under a road with larger culverts. These allowed the tide to flood the marsh again and together with the reopening of channels that had filled with silt and *Phragmites australis*, provided the conditions for saltmarsh re-creation. This method resulted in reduced breeding opportunities for mosquitoes and a substantial decrease in *Phragmites* height and vigour. Restoration is gradual, taking many decades following the re-establishment of the tidal regime. Details of this and other restoration schemes in the Rhode Island area, appear on the 'Habitat Restoration Portal' (see http://www.edc.uri.edu/restoration/index.htm).

A review of the methods of Regulated Tidal Exchange, (RTE), including those in the USA, describes the use of pipes, sluices or tide-gates to allow regulated tidal flushing by seawater to create saline/brackish habitats behind sea walls. Techniques include:

- Open culvert, with no tidal flap through the sea wall;
- Culverts with manually operated flaps to let water through into an impoundment at high tide;
- Self-regulating tide-gates;
- Electronically controlled and operated tide-gates (Lamberth & Haycock 2002).

Three factors appear to retard the period for vegetation establishment:

1. Limited suspended sediment;
2. Erosion of deposited sediments by wind-generated waves;
3. Restricted tidal exchange (Williams & Orr 2002).

## 6.3.2 The German Baltic Coast

There are also several examples of re-integration on the German Baltic coast, including two large managed realignment schemes creating 700 ha of tidal land, in the federal state of Mecklenburg-Vorpommern. In a pilot project near Greifswald on the coast of the Greifswalder Bodden, natural periodic flooding returned, following breaching of the front line of defence in 1993. By 1998, typical saltmarsh vegetation covered 75% of the 350 ha, showing that large-scale restoration saltmarsh is possible (Bernhardt & Koch 2003). In the Baltic Sea (as in the UK), the defences were already in a relatively dilapidated state and re-integration was a more cost-effective option than in the German part of the southern North Sea, where sea defence was one of the principal issues (Rupp & Nicholls 2002; Bakker et al. 2002).

## 6.4 Re-creating Saltmarsh from Agricultural Land in England, 'Abandonment' or 'Managed Realignment'

Throughout the world substantial areas of agricultural land have been 'won' from the sea through the enclosure and drainage of tidal saltmarshes. Some of the larger sites surround the southern North Sea. It is not surprising that in an area where the land is sinking relative to sea level, resulting in a 'coastal squeeze' that the economic justification for 'protecting the land at all costs' has receded (Section 3.5.2).

Two principal approaches are involved:

1. Abandonment, where the 'summer dykes', sea walls or other structures are simply left to deteriorate, allowing the sea to reinvade, often in response to major storm events;
2. Realignment, involving active intervention to re-establish tidal movement by removing part of the enclosing sea walls.

## *6.4.1 Abandonment*

Abandonment is a haphazard process resulting from changing economic circumstances and extreme storm events. The situation in south-east England is illustrative of the process. Saltmarsh enclosure has created many thousand hectares of high quality farmland (Section 2.4.2). The total area of coastal grazing marsh (derived from enclosed saltmarsh) in Essex alone is over 3,000 ha (Boorman & Ranwell 1977). Much of this low-lying land is at risk from flooding, as happened in the 'Great Tide' of 1953 (Grieve 1959; Figure 49).

The 1953 flood followed the pattern of other disasters in the twelfth, thirteenth and fourteenth centuries. These continued into the sixteenth century, but despite the loss of human life, cattle, sheep, and damage to grazing land 'marshland once reclaimed was obstinately held on to' (Grieve 1959). Regular flooding involving breaches to sea defences, occurs on a regular basis when storms arising from the northeast coincide with spring tides. Major floods of 'unprecedented levels' occur approximately every 55 years, at least in Essex (Burd 1992). The outcome of these, and the most recent 1953 storm surge, made it uneconomic to repair sea walls in front of some areas of flooded land. Within these areas, the tide re-established its normal cycle, eventually re-creating saltmarsh.

A number of 'abandoned' areas of former saltmarsh were identified from historical records, maps and aerial photographs (Burd 1992). The following list shows the estuaries involved and the dates of abandonment of individual areas:

- Deben Estuary: 1953;
- Hamford Water: 1338, 1874, 1896, 1921, 1953;
- Colne Estuary: 1897, 1921, 1945, 1953;
- Blackwater Estuary: 1897, 1945;
- Crouch Estuary: 1897, 1953;
- Thames Estuary: 1874 (Burd 1994).

1897 and 1953 were the principal years when many of the breaches (approximately 50%) occurred (Wolters et al. 2005). The quality of these 'naturally' re-created saltmarshes varies depending on environmental conditions (tides, sediment supply, and time). Century-old 'reactivated' saltmarshes showed no discernable difference

## 6.4 Agricultural Land in England

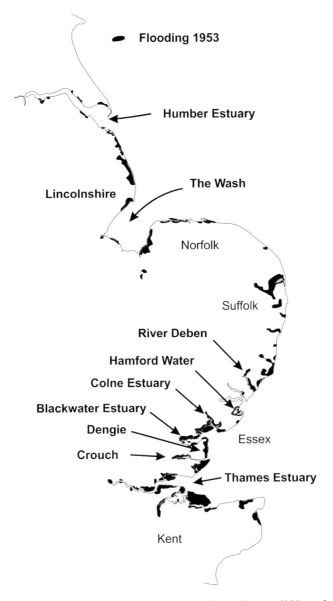

**Figure 49** Approximate locations of the 60,000 ha of land along 1,600 km of the eastern coastline of England, flooded by the 'Great Tide' in 1953. Over 300 people lost their lives on land, 24,000 homes flooded and 40,000 people evacuated. Information derived from a number of web sites, such as the Environment Agency (see http://www.environment-agency.gov.uk) and Risk Management Solutions (Anon 2003). The figure shows sites in the east of England mentioned in the text

**Figure 50** Former enclosed saltmarsh developed for agriculture, 'naturally' restored following the 1953 storm surge. Note the dead hedgerow trees at the upper margins of the saltmarsh

from natural ones, in terms of their vegetation. These restored habitats may have different physical characteristics and environmental functions to those of nearby unenclosed saltmarshes. For example, restored areas consist of consolidated sediments (due to compaction through drainage), and have a different creek pattern to unenclosed saltmarshes (Atkinson et al. 2001). An examination of a number of naturally restored saltmarshes showed similar results. In addition to elevation, drainage was important in determining creek density (Crooks et al. 2002).

The later breaches, such as those of 1953, also developed into typical vegetation with *Aster tripolium* dominated communities (Burd 1994). On the River Deben, the vegetation associated with inundated agricultural land, includes *Atriplex portulacoides*, *Puccinellia maritima* and *Aster tripolium* (Figure 50).

In 1990, a breach in a brackish part of the Scheldt Estuary, the Netherlands occurred returning the Sieperda polder to the influence of tidal waters. The breach remained open and a succession to brackish saltmarsh and mudflat took place rapidly (Eertman et al. 2002). This example lies somewhere in between an approach which essentially says, 'leave things alone' to one involving active intervention to 'realign' the coast (Bakker et al. 2002).

## 6.4.2 Realignment

Realignment represents a deliberate attempt to re-create saltmarsh on enclosed tidal land. In the UK, this process developed in response to recognition that, in the south-east where land is sinking because of isostatic adjustment, continuing to raise sea walls against rising sea levels was unsustainable in some areas. The method adopted was called 'managed retreat', 'set back' and more recently 'managed realignment'. In Continental Europe the term, 're-integration' (with the sea) was used in Germany, although this included Regulated Tidal Exchange (Section 6.3.2), and 'depoldering' in the Netherlands.

The methods employed are, in principle, relatively straight forward, involving an engineered breach in the existing seaward sea wall. This may include building a new sea defence, or reinforcing an existing one landward of the breach. Two important features help to secure successful realignment, namely:

- An appropriate elevation in relation to the tidal regime. The limit of terrestrial vegetation (normally reached by species of *Spartina*) defines the lower vertical range for saltmarsh; transitions to non-saline vegetation defines the upper limit (Hough et al. 1999);
- A degree of protection from wave exposure enables pioneer plants to become established.

It is also essential to have a suitable and adequate supply of sediment. The factors important for the establishment of saltmarsh under 'natural' conditions (Chapter 3) are also those needed for realignment.

Some or all of the following considerations help to determine the engineering requirements:

- **Nature of the landward transition**. This is likely to be rising ground, a previous sea wall or new embankment. Sites where there is rising ground and no urban development require least intervention by way of new defences. On open coasts, estuary mouths and areas with urban populations, the artificial realigned 'retreat line' may need to be substantial. Another factor which helps determine the investment is the width of the available realignment area;
- **Crest height**. Wave modelling, overtopping information, vegetation type and likely extreme events will form part of the analysis to decide on the height of any new or improved embankment;
- **Bank construction**. Earth embankments derived from locally available material, provide the most cost-effective and environmentally sound option;
- **Natural sea defences**. At some sites, shingle ridges or sand dunes may provide a natural defence. Allowing these to develop without intervention, in response to prevailing wave, tidal and storm conditions, will result in breaches and/or landward movement of the structures;
- **Changing land level**. Importing material to raise the levels within the realignment area, or redistribution to create lower levels, will help create conditions suitable for saltmarsh development or mudflats, respectively;
- **Creeks**. Excavating former creeks help improve tidal flows, sediment distribution and reinstatement of channels;
- **Surface roughness**. Breaking up the surface mechanically will aid sedimentation and vegetation establishment on land, consolidated as a result of cultivation;
- **Vegetation**. Plant material brought in by the re-established tidal regime is likely to provide the best opportunities for recreating 'natural' vegetation;

- **Breaches**. Location and size of the breaches help to establish former tidal conditions and creek patterns. In some instances, the breach may involve a reduction in height of the seaward bank to create a spillway (Leggett et al. 2004).

Examples from the UK range from the first experimental site at Northey Island (Section 3.4.2), which was only 0.8 ha to the 400 ha Alkborough site, on the Humber Estuary. To date there are 16 realignment sites in the UK, 9 in the Netherlands, 2 in Belgium and 9 in Germany either completed or planned. In the Netherlands, four of the schemes were abandoned due to local opposition and one of the two schemes in Belgium (Wolters et al. 2005). In the UK, the majority combine deliberate breaches with strengthening of a defence landward. Examples include the Freiston scheme (Section 3.5.3), Tollesbury and Alkborough.

Tollesbury lies within the Blackwater Estuary on the Essex Coast and is one of a number of sites included in a European Union LIFE Project 'Living with the Sea' (Worrall 2005). It is illustrative of an experimental approach designed to test the extent to which habitats, (saltmarsh and tidal flats) develop and support invertebrate and bird species. Situated on the north shore of the estuary, and originally embanked in the late eighteenth century, it covers an area of some 21 ha. It is one of the first 'relatively' large-scale (for the UK) experimental managed realignment schemes.

Although the lower mudflat areas had not colonised after 7 years of tidal influence, monitoring data showed that considerable accretion had taken place. Accretion rates were high averaging 23 mm per year, ranging from 8 mm to 258 mm. Approximately, 6 ha of the 21 ha site had been colonised by saltmarsh vegetation. By October 2002 (Figure 51) the vegetation consisted mainly of

**Figure 51** Developing saltmarsh in 2002, following a deliberate breach in the sea wall in 1995, Tollesbury, Blackwater Estuary, Essex

*Salicornia europaea* with *Puccinellia maritima* and *Atriplex portulacoides*. The last species was restricted to the highest elevations near the foot of the new sea wall.

Invertebrate colonisation of the site was rapid, with 14 species recorded after only two months of tidal inundation and between 18 and 19 species, thereafter with the number remaining constant. The most abundant species was the Mud Snail (*Hydrobia ulvae*) recorded throughout the site by 2001. Garbutt et al. (2006) provides a review of the results of the vegetation, sedimentation and invertebrate monitoring work. It discusses the implications for other managed realignment schemes in the UK. Standardised bird monitoring showed that after 5 years the range of species using the site was similar, but not identical to nearby 'natural' areas. Four years after inundation, the bird populations were still evolving (Atkinson et al. 2004).

Alkborough is a larger and more ambitious scheme. It is designed to improve flood protection and at the same time help re-create tidal habitats including saltmarsh. A combination of engineering works will create a 20-m wide breach in the existing sea wall as well as 1,500 m of lowered embankment (spillway) in an area of high quality agricultural land (Figure 52). Improved hard defences will protect the assets around the edge of the site. Some of the land will become mudflat and saltmarsh. On higher ground, transitions to reed bed will complete the habitat restoration. Inundation of the highest parts of the site will only take place on occasions when particularly high tides overtop the spillway. At other times, grazing will take place on this higher ground.

**Figure 52** Looking over the Alkborough realignment site. Much of the area was high quality agricultural land, when this picture was taken in 2004. The River Humber appears in the distance

Whilst the engineering options for restoration are relatively easy to define, their implementation is much more complicated. Legislation in the coastal zone is generally highly complex. In the UK, for example, some or all of the following laws apply:

- Town and Country Planning Act;
- Coast Protection Act;

- Food and Environmental Protection Act;
- Environment Impact Assessment Regulations;
- Land Drainage Act;
- Water Resources Act;
- Flood Defence (Land Drainage) Bylaws and Sea Defence Bylaws;
- Highways Acts.

Other national legislation and European Directives relating to environmental issues also require consideration, including those relating to:

- Nature conservation;
- Landscape and historic environment;
- Agricultural land use;
- Recreational use of land or sea;
- Navigation;
- Water resources or quality.

Taken together with the need to ensure local support, the process of developing proposals is a complex and often time-consuming activity.

## 6.5 Conclusions

Evidence from around the world shows that saltmarsh can be restored. The methods are many and various, as shown above. A key question arises in relation to any restoration project – how successful are they in fulfilling the original aims? Chapter 10 discusses this further. Zedler (2000) provides a useful summary using examples derived from California. The following paragraphs introduce some of the sources of information available on the Web, which relate to coastal wetland restoration, including saltmarshes.

In the USA the National Oceanic and Atmospheric Administration (NOAA) Coastal Services Center is a useful starting point for information on coastal restoration projects, methods and rational for wetlands, including saltmarshes. It also includes a section on 'Innovative and Successful Monitoring and Adaptive Management Approaches', (see http://www.csc.noaa.gov/coastal/). These, together with sites concerned with the successes and failures of habitat restoration, contribute to the overall aim within the USA to 'Restore America's Estuaries' (see http://www.estuaries.org/). This initiative has spawned numerous studies, case studies and large-scale restoration. The *Ramsar Convention on Wetlands* includes 'Principles and guidelines for wetland restoration' (see http://ramsar.org/key_guide_restoration_e.htm).

The Internet provides a vast array of more specific information, with many individual site reports. The following web sites provide more general advice on saltmarsh restoration and monitoring:

- The Gulf of Maine Council on the Marine Environment provides guidance on site selection, setting goals for restoration, baseline data needs, funding opportunities,

## 6.5 Conclusions

regulations, restoration techniques and design considerations for saltmarsh restoration (see http://restoration.gulfofmaine.org/projectplanning/saltmarsh.php). Monitoring protocols are outlined in Neckles et al. (2002);
- CICEET: The Cooperative Institute for Coastal and Estuarine Environmental Technology includes a study on saltmarshes – log on to http://ciceet.unh.edu/ and search on 'saltmarsh' (Rogers 2005). The work aims to provide a tool for monitoring saltmarsh, following their restoration through self-regulating tide-gates in the Gulf of Maine;
- Saltmarsh Restoration Project in Nova Scotia (see http://www.ecologyaction.ca/EAC_WEB_1/PROJECTS/salt_marsh/SaltMarsh.shtml) describes the use of replacement and/or enlargement of tidal culverts;
- Environment Canada provide details of saltmarsh conservation and restoration (see http://www.atl.ec.gc.ca/wildlife/salt_marsh/restoration_e.html);
- In the UK the University of East Anglia hosts a site 'Restoration and Creation of Saltmarshes and other Intertidal Habitats', including an extensive bibliography of published work, Alastair Grant, Centre for Ecology, Evolution and Conservation, (see http://www.uea.ac.uk/~e130/Saltmarsh.htm). It provides links to information on some of the main saltmarsh restoration sites in the UK. The site includes work undertaken for English Nature (that part of Natural England dealing with nature conservation issues) and reported in Atkinson et al. (2001).

# Chapter 7
# Vegetation States

## Trends and Trade-Offs

## 7.1 Introduction

The physical conditions determine whether saltmarsh develops or not. Its lateral extent and vertical accretion rates depend on the establishment and growth of vegetation. Under natural conditions the tides, sediment dynamics drainage and exposure to waves influence the way the vegetation responds. Climatic variation also plays a part in determining the type of plant communities developing in different geographical locations (Section 4.5.1).

Grazing, especially by domestic stock, (mainly cattle and sheep) fundamentally changes both the 'State' of the saltmarsh and the 'values' associated with it (Sections 4.4 and 4.5 respectively). This in turn has a profound affect on the biological condition of the saltmarsh. The nature of the vegetation (structure and height of the sward) also affects its resilience and ability to dampen wave energy (Section 4.3.2). However, this chapter is primarily concerned with the 'quality' of the vegetation from a nature conservation perspective. These involve issues, such as the presence of specialist plant and animal communities, vegetation complexity and structure, and the way they change under different grazing regimes.

## 7.2 Mechanisms for Change – Native Species

Natural herbivores graze saltmarsh and can change the structure of the vegetation and influence its biodiversity. However, the introduction of domestic grazing animals (notably sheep and cattle) has, in most situations, a greater influence. This section provides a summary of the way in which the different patterns of grazing, by both natural and introduced animals, influence the type of vegetation and its associated fauna.

## 7.2.1 Native Waterfowl in Europe

Herbivorous ducks and geese graze saltmarsh. Individual birds often seek out the more palatable and nutritious plants. For example, a study of the diet and habitat use of moulting Greylag Geese (*Anser anser*) on the island of Saltholm, between Denmark and Sweden, showed that the geese fed almost exclusively on *Puccinellia maritima* during the main moult period. The plant species exhibited the highest levels of protein of any of the grass species studied (Fox et al. 1998). Despite its limited spatial distribution, the geese seek out the plant, which helps to sustain and in some cases extend, the spread of the species.

These, and other waterfowl, can occur in large numbers at individual sites. For example in the Wadden Sea, in the southern North Sea, the total number of wintering ducks and geese has been as high as 1.66 million. Included within this number are the Barnacle Goose (*Branta leucopsis*) and Dark-bellied Brent Goose (*Branta bernicla bernicla*), which have exceeded a quarter of a million birds. Numbers of Wigeon, another herbivorous species, reached more than 330,000 (Blew & Südbeck 2005). These numbers represent a massive increase in the populations in recent years, with the Barnacle Goose up from about 20,000 individuals in 1960 to over 250,000 in 1996, since when it has stabilised. Similarly, the population of the Dark-bellied Brent Goose increased from 17,000 individuals in the 1950s to more than 250,000 individuals in 1995. Additionally, in the beginning of the 1970s, the last species enlarged its breeding range from northern Russia to the coasts along the Baltic Sea (Laursen 2002).

Other species show a similar trend. A comparison of the populations of 13 species of avian herbivores, showed expanding populations over the last 50 years. Species protection and the use of artificial nitrogen fertilisers, appear to have increased the winter carrying capacity. Although populations of the smaller species, such as ducks, are either stable or have peaked and are now in decline, numbers of larger herbivores (geese and swans) continue to increase (Van Eerden et al. 2005).

Despite these large numbers in Europe, there are no references to substantial goose damage to saltmarsh. Most of the larger populations of Dark-bellied Brent and Barnacle Goose, for example in the Wadden Sea, graze agricultural crops throughout the winter, only returning to saltmarshes in late spring. This represents a change in behaviour resulting from the increase in use of fertilisers on agricultural land, making the swards more palatable to these species (Van Eerden et al. 2005). In the UK, the same species also frequent agricultural land, prompting economic analysis of alternative approaches to controlling this use, including:

- Doing nothing;
- Culling;
- Paying compensation;
- Providing alternative feeding areas (Vickery et al. 1994).

The Greater White-fronted Goose (*Anser albifrons albifrons*) also causes problems for farmers and in the Netherlands and Germany, annual compensation is paid. Due to the smaller, more dispersed population and shorter residency of the species in the Britain, similar problems do not occur for this species (Hearn 2004). In Denmark, migrating Pink-footed Geese (*Anser brachyrhynchus*) also cause conflict with farmers (Jepsen 1991). Increasing goose populations, more generally, cause decreasing yields in grass and cereal production, although the effects vary considerably (Patterson 1991).

Amongst the options for improving the grazing opportunities for geese and reducing the impact on agricultural land, is increasing the grazing pressure on saltmarshes to create shorter swards that are more palatable. This has implications for the values associated with ungrazed or lightly grazed saltmarshes, as is discussed in Chapter 8.

As in Europe, numbers of geese have increased rapidly in North America. The population of Lesser Snow Goose (*Chen caerulescens caerulescens*) reached nearly six million individuals, before the introduction of control measures. The impact of these numbers on saltmarshes in their breeding grounds in Arctic Canada is significant. In particular, they can alter the composition and structure of the vegetation to the detriment of the habitat and its associated plant and animal species, (dealt with in more detail in Section 7.4).

### 7.2.2 Mammals

In Europe, in a temperate saltmarsh in the Netherlands, Brown hares (*Lepus europaeus*) retard vegetation succession and facilitate grazing by Brent geese. Winter grazing by hares prevented the shrub *Atriplex portulacoides* from spreading in younger parts of a saltmarsh, and by reducing the number of dead *Seriphidium maritimum* (*Artemisia*) stems in grassy swards. Both pant species hamper grazing by geese and along with grazing by domestic stock, are important factors facilitating feeding conditions for Brent geese along the north-west European coast (Van der Wal et al. 2000).

In the upper reaches of Chesapeake Bay, in the USA (Choptank River Maryland) Muskrat graze the Narrow-leaved Cattail (*Typha angustifolia*). The total loss of surface vegetation 'Eatouts' have led to the conversion of a solid mat of rhizomes, to an unconsolidated, anoxic, organic substrate. The Muskrat 'eatout' also led to the loss of peat structure (including airspace), lowering the marsh surface by 2–6 inches (Garbisch 1994). This activity creates open spaces in the Cattail stands, decreasing Cattail density, diversifying habitat and allowing other species to become established.

### 7.2.3 Introduced Nutria in North America

In North America, Nutria (*Myocastor coypus*) introduced for the fur trade, have had a significant adverse impact on saltmarshes and other coastal wetlands. Their powerful jaws and forefeet, which they use to dig under the marsh surface and feed directly on the root mat, cause pitting in the marsh. They also leave deep channels through which they swim.

The population has increased dramatically. For example, estimates on the 10,000 acres of the Blackwater National Wildlife Refuge, grew from about 250 animals in 1968, to between 35,000 and 50,000 animals in 2003. The Nutria appear to be a primary force in accelerating marsh loss in the Blackwater basin, causing the destruction of the vegetative root mat which holds the marsh together. These problems led to the establishment of the Chesapeake Bay Nutria Working Group, which set a goal in 2003 of eradicating the species from the Chesapeake watershed by 2009 (Chesapeake Bay Nutria Working Group 2003).

Other studies in North America suggest Nutria eat large areas of vegetation in many coastal regions of south-central and south-east Louisiana. Vegetation removal contributes to permanent loss of vegetated wetlands and may have played a significant role historically, in changing plant species composition throughout the coast. Only a small fraction of damage sites were observed to have recovered since initial surveys in 1993 (Kinler et al. 1998). In 1998, a project called 'Nutria Harvest for Wetland Restoration Demonstration (LA-03a)' began. The aim was to eradicate nutria by 2009, with the initial programme completed in 2003 (see http://lacoast.gov/projects/overview.asp?statenumber=LA%2D03a). After this, a further Coastwide project, agreed by the Louisiana Coastal Wetlands Conservation and Restoration Task Force, the 'Nutria Control Program (LA-03b)' came into force, reflecting the wide scale of the problems associated with this introduced animal.

'The chance of restoring or even slowing the degradation of coastal marshes in Louisiana will be hampered considerably without sustained reduction of Nutria populations'. Taken from the Louisiana Department of Wildlife and Fisheries web site which is devoted to Nutria (see http://www.nutria.com/site.php).

The same species (*Myocastor coypus*), called Coypu in the UK, caused similar but less extensive problems in south-east England. These affected inland waterways in East Anglia rather than saltmarshes. Early intervention and natural losses caused by cold winters, helped an eradication programme funded by the UK Ministry of Agriculture, Fisheries and Food. Introduced in the 1920s, successful control of the population had taken place by 1986 (Gosling & Baker 1989). The absence of any evidence of Coypu, 20 months later resulted in the discontinuation of the eradication programme in 1989 (Fasham & Trumper 2001).

### 7.2.4 Snails and other Invertebrates

The food web for saltmarshes includes primary consumers feeding directly on the vegetation. Whilst geese, other avian herbivores, and some mammals, are the most obvious, invertebrates also graze the plants. Amongst these, the snails *Hydrobia* in Europe and the dominant marsh grazer on saltmarshes in America, the Saltmarsh Periwinkle (*Littoraria irrorata*), which grows up to 24 mm may be significant (Silliman & Zieman 2001; Silliman & Bertness 2002). Overharvesting of snail predators (e.g. blue crabs *Callinectes sapidus*) may be important in contributing to 'die back' of saltmarshes across south-eastern USA (Silliman & Bertness 2002).

This reflects a debate in the UK about the efficacy of managed realignment in the face of grazing by the Ragworm (*Nereis diversicolor*). Research suggests that this infaunal polychaete causes the loss of pioneer zone plants, increases sediment instability and exacerbates the erosion of saltmarsh creeks, a major cause of saltmarsh erosion in south-east England (Paramor & Hughes 2004; Hughes & Paramor 2004). Whilst there may be some effects, claims that accidental abandonment of defences have not resulted in saltmarsh establishment (Hughes & Paramor 2004) are not borne out by experience (see Section 6.4.1 above). Further they claim, 'many realignment areas are unlikely to develop vegetation', a conclusion not supported by a review of 37 abandoned and deliberately realigned sites in Europe (Wolters et al. 2005). Another argument suggests that tidal flats with *Nereis diversicolor* are too low for the establishment of pioneer saltmarsh plants (Dijkema, pers. comm.).

Part of the debate rests on the use of the phrase 'coastal squeeze'. Described in Section 3.5.2, this is a representation of what is happening, rather than an explanation of cause and effect. The loss of saltmarsh in south-east England results from a variety of causes, including lack of sediment, embankment, climate change and sea-level rise (Doody 2004).

## 7.3 Mechanisms for Change – Grazing by Domestic Livestock

Historically, saltmarshes provided pasture for domestic livestock in many parts of the world (Section 2.1.1). The density of stock ranges from those sites that are historically ungrazed to heavily sheep-grazed saltmarshes. Grazing at moderate to high levels continues today in areas such as western Ireland, north and western Great Britain, and in parts of the Wadden Sea in the Netherlands, Germany and Denmark. The number and type of stock and the duration of grazing are key factors in determining the state of the vegetation (described in Section 4.4) and biological values associated with it, including the animals that inhabit the saltmarsh (described in Section 4.5). When considering change in grazing patterns, therefore, the principal considerations are the effect on the vegetation and the impact on the associated animals.

### 7.3.1 Changes in Plant Communities

Grazing affects saltmarsh vegetation in complex ways, with individual species responding differently. A grazing experiment in the Wadden Sea showed that, at 'moderate' stocking levels, most plant species were positively associated with grazing. *Puccinellia maritima* and *Festuca rubra* showed a positive response and together with Saltmarsh Rush (*Juncus gerardii*) favoured grazed sites. Grazing negatively influenced *Atriplex portulacoides* and *Elytrigia atherica*, and communities dominated by these

## 7.3 Mechanisms for Change – Grazing by Domestic Livestock

species and Sea Wormwood (*Seriphidium maritimum*) were restricted to lightly grazed and ungrazed saltmarsh. Species richness also increased with elevation, and was 1.5–2.0 × higher in the grazed than ungrazed saltmarsh (Bos et al. 2002).

In a study of grazed and ungrazed saltmarsh in France, under grazed conditions *Puccinellia maritima* dominated a short turf. With cessation of grazing, *Atriplex portulacoides* replaced *P. maritima* in the well-drained lower marsh, surviving in the mid-marsh, possibly due to fine sediments and poor drainage. Sea Couch *Elymus pungens* (*Elytrigia atherica*) cover was initially limited, but after 6 years began to increase and appeared likely to become dominant in the future. Annual species, such as *Salicornia europaea* and *Suaeda maritima*, decreased in saltmarshes sheep-grazed between February and June, although this had no effect on *P. maritima* (Tessier et al. 2003). In Denmark, enclosure experiments showed a similar effect, and after 35 years the vegetation originally dominated by *P. maritima* and *S. europaea* became dominated by *Atriplex portulacoides* in the absence of grazing. The grazed saltmarsh remained dominated by *P. maritima* and *S. europaea* (Jensen 1985).

After 9 years, removal of grazing from an intensively grazed saltmarsh in Germany resulted in the following changes in species composition:

- Cover of *Aster tripolium*, *Atriplex portulacoides* and to a lesser extent, *Seriphidium maritimum* and *Elytrigia atherica* increased;
- Cover of *Salicornia europaea* decreased;
- A strong increase in the first years *Aster tripolium* was followed 2–6 years after abandonment by a decrease;
- *Puccinellia maritima* was replaced by *Festuca rubra* in the mid saltmarsh zone;
- Cover of *Puccinellia*, *Festuca*, *Suaeda maritima*, *Glaux maritima* and *Salicornia* showed strong fluctuations, possibly due to differences in weather conditions and inundation frequency.

The conclusion after 9 years was that 'cessation of grazing did not lead to large-scale dominance of single plant species' (Schröder et al. 2002). However, this may be too short a timescale! Another study in the Wadden Sea, on high saltmarsh in the island of Hallig Langeness, Germany, showed that lower grazing intensities, or the removal of grazing altogether, resulted in *Elymus* spp. [presumably Sea Couch (*Elytrigia atherica*)] becoming dominant within saltmarshes with 'summer dikes'. At the same time, there was a strong decline in species richness (Kleyer et al. 2003). The same study gives an insight into the relationship between grazing, groundwater depth, soil salinity and plant diversity. Cattle grazing at moderate stocking rates of 0.6 livestock units per ha (equivalent to approximately 3 sheep per ha) were optimal for plant diversity conservation. The relative proportion of each depends on the depth of the groundwater, with *Elymus* spp. dominating ungrazed sites with lower ground water. Succession in areas with higher water levels is restricted and low-growing halophytes dominate (Kleyer et al. 2003).

Moderate stock levels also provide the best opportunities for micropatterns in *Festuca rubra* dominated saltmarsh to develop. Experiments using five different stocking rates (0, 1.5, 3.0, 4.5 and 10 sheep per ha) showed that at the highest and lowest stocking levels micropatterns did not develop. Moderate rates, especially 3 sheep per ha, showed the greatest spatial variation. Moderation of these changes by rainfall and sediment deposition resulted in very subtle interactions (Berg et al. 1997).

A grazing study in the German Wadden Sea (Leybucht, Niedersachsen) further illustrates some of the trade-offs with changing grazing pressure. Three stocking rates, namely 0.5, 1.0 and 2.0 cattle per ha, were compared with an abandoned area between 1980 and 1988. The canopy height decreased with increasing stocking rate, as did sedimentation. Grazing also reduced many of the species living in, or on upper parts of, the vegetation as well as those sensitive to trampling. The community structure changed from a food web dominated by plant-feeding animals to one dominated by detritus feeders. Over the relatively short 9-year period of the study invertebrate population densities, species-richness and community diversity increased with the cessation of grazing (Andresen et al. 1990).

However, *Elytrigia atherica* became dominant in the higher saltmarsh in the abandoned site, suggesting that this might become dominant over the whole site in the absence of grazing (see below). It is perhaps too early to make sweeping conclusions from these studies, although they all show that changes to the grazing regime are important to the plant communities and particularly the dominance of key species. The second two studies are relatively short (8 and 9 years respectively) and longer periods are required to adequately reflect the changes that can occur. The review of grazing experiments with different stocking rates of cattle on saltmarshes in the Netherlands and Germany suggests that long-term (over 20 years) *Elytrigia atherica* can become dominant (Bakker et al. 2003). With this dominance the range of plant communities decreases. Over shorter timescales <10 years, experiments in the Wadden Sea show no dominance by *Elytrigia atherica*.

Another important consideration is the type of grazing animal. Sheep and cattle have very different methods of grazing. The former, selectively bite individual plants (sometimes selecting flowers only) and will graze close to the soil surface. The latter, on the other hand, tear the vegetation leaving more of the above ground parts extant. This tends to make sheep-grazed saltmarshes uniform in their structure, by comparison with those that are cattle-grazed. This has important consequences for several of the key groups of animals utilising the saltmarsh.

Grazing thus has a complex and significant influence, not only on the distribution of halophytes within the saltmarsh, but also on the diversity of plant species. These can be both positive and negative, in relation to plant diversity (Ungar 1998). Overall, however, the general conclusion that plant species diversity is lowest in heavily grazed and abandoned, formerly grazed saltmarshes, remains true. These differences are most obvious on sheep-grazed sites, as described in Chapter 4.

### 7.3.2 Changes in Rare Species

The direct effects of grazing restrict the number of plants occurring in many saltmarshes in Europe. This not only relates to the consumption of the plants themselves, but also to the impact of trampling. Some species such as *Atriplex portulacoides*, are especially susceptible (Adam 1990). The general conclusion that heavy grazing pressure reduces species diversity also appears to be true for some of the rarer saltmarsh plants. There is no detailed research work, and this assertion is based largely on the absence of species from many heavily grazed sites. Species such as Sea Lavender (*Limonium* spp.) would probably occur much more widely in saltmarshes of north-west England in the absence of grazing (Adam 1990).

Although not directly related to grazing management, some rare species depend on transitions between different stages in saltmarsh development for their survival. Transitions to terrestrial vegetation are most obvious where a sand dune or shingle feature encloses the saltmarsh. These tend to occur in barrier island complexes such as those of the North Norfolk coast. Transitions to reed bed, tidal woodland and scrub in estuaries are particularly important. Chapter 2 shows how human use, notably enclosure of estuarine habitats, has modified or destroyed many transitions. In the international Wadden Sea, almost all have been destroyed (Dijkema pers. comm.).

As the list below shows, the rarer species tend to be associated with the upper margins of the saltmarsh and transitions to other terrestrial habitats (Table 15). All of these species are restricted to the saltmarshes of south-east and southern England, where grazing is limited.

Note that some of the most visually distinct saltmarshes, such as those dominated by *Armeria maritima*, appear to tolerate grazing. Here they form extensive communities on heavily sheep-grazed saltmarshes, as for example, in north-west Scotland (Figure 7). Transitions at these sites from upper saltmarsh with the northern *Blymus rufus* to fen with Yellow Flag (*Iris pseudacorus*) can be rich in species.

Table 15 Rare (R) and scarce (S) saltmarsh species in Great Britain (The definition of rarity depends on the recorded presence of the species in $10 km^2$. Species, which occur in 15 or fewer $10 km^2$, are 'nationally rare' and those occurring in 16–100 $km^2$ 'nationally scarce').

| Species | Distribution | $10 km^2$ | Status | Habitat |
|---|---|---|---|---|
| *Limonium bellidifolium* | eastern | 5 | R | High level sand/mud |
| *Spartina alterniflora* | south | 1 | R | Low level mud flat |
| *Atriplex pedunculata*\* | east | 1 | R | Grazing marsh |
| *Atriplex longipes* | scattered | 7 | R | Upper marsh |
| *Suaeda vera* | south | 30 | S | Drift line/shingle |
| *Bupleurum tenuissimum* | south | 58 | S | Upper marsh/grazing marsh/sea walls |
| *Limonium binervosum* | south | 70 | S | Saltmarsh/dune transition |
| *Spartina maritima* | south-east | 25 | S | Low marsh |
| *Frankenia laevis* | south-east | 25 | S | Upper marsh/sandy flats |
| *Salicornia pusilla* | south-east | 32 | S | Low marsh |
| *Salicornia perennis* | south-east | 37 | S | Low marsh |
| *Puccinellia fasciculata* | south-east | 52 | S | Open saline areas above extreme high-water spring tides |
| *Limonium humile* | south | 59 | S | Upper/mid-marsh |

\* Species now confined to grazing marsh in GB (one site only), though formerly present on upper saltmarsh

Grazing may be positive for some rare species. Removal of rabbit grazing on a saltmarsh fringing Scolt Head Island on the North Norfolk Coast, south-east England, seems to have resulted in the replacement of the rare, Matted Sea-lavender (*Limonium bellidifolium*) with *Puccinellia maritima* and *Atriplex portulacoides* (Chapman 1960).

## 7.3.3 Grazing and Breeding Birds

The type of saltmarsh favoured by breeding birds is dependent on their nesting preferences. Ground-nesting species have different strategies, some use camouflage, nesting in the open, whilst others hide the eggs in a tussock or other feature. The Oystercatcher is one such species adopting the former strategy, the Redshank the latter. The value for breeding passerines also depends on there being a good structural and species diversity in the transition zones. Breeding birds represent a major conservation interest on many saltmarshes, which can be the only suitable nesting habitat in areas of intensive agriculture. The Table 16 gives an indication of the nature of the habitat required, the grazing pressure and the density recorded for some species.

**Table 16** Habitat, geographical location and density of breeding birds on saltmarshes in Great Britain (Cadbury et al. 1987)

| Species | Habitat | Distribution | Average Nos. |
|---|---|---|---|
| Redshank (*Tringa totanus*) | Tussocky, moderately grazed mid-high marsh | 50% East Anglia, 25% north-west England | Up to 100 pairs per km$^2$ (The Wash) |
| Oystercatcher (*Haematopus ostralegus*) | Moderately to heavily grazed high marsh | Highest densities in the north-west | 72 pairs per km$^2$ (The Solway) |
| Lapwing (*Vanellus vanellus*) | Grazed grassland on high marsh | western England & eastern Scotland | 20 per km$^2$ |
| Black-headed Gull (*Larus ridibundus*) | Mounds on marsh; also in *Spartina* islands | Scattered large colonies | 20,000 in 1 colony (Ribble Estuary) |
| Skylark (*Alauda arvensis*) | Upper grassy marsh | Widely distributed | Up to 105 pairs per km$^2$ |
| Reed Bunting (*Emberiza schoeniclus*) | Edge of marsh with tall vegetation, moderate to light grazing | Widely distributed | Up to 80 pairs per km$^2$ |
| Meadow Pipit (*Anthus pratensis*) | Upper grassy marsh, moderate to light grazing | Widely distributed | Up to 76 pairs per km$^2$ |

Moving from a lightly or moderately grazed saltmarsh to a heavily grazed one, is likely to influence the species of bird nesting in the habitat. Breeding Redshank show the nature of the changes that can occur. In the UK, saltmarshes now support a substantial proportion of the population, because of loses through drainage of tussocky wet grasslands, their preferred habitat, elsewhere. Studies on the Wash, eastern England, show a number of important factors that influence the value of the saltmarsh for nesting Redshank, namely:

1. The structure of the saltmarsh is of paramount importance. The species seeks out grassy tussocks in which to hide the nest;
2. There is no obvious correlation between habitat diversity and breeding as implied by other studies;
3. *Elytrigia atherica* expansion is positively correlated with redshank breeding on 'heavily' grazed sites and negatively on ungrazed areas;
4. A marked decrease in breeding pairs occurs when grazing patterns change from ungrazed/lightly grazed to moderately/heavily grazed;
5. 'Moderate' grazing by cattle helps to create the optimum habitat with >25% cover, which provides a good nesting habitat (Norris et al. 1997, 1998).

A key pattern emerges, which relates to the structure of the saltmarsh. *Elytrigia atherica* tends to dominate many areas of the Wash saltmarsh, because of the reduction in grazing pressure over the last few decades. This abandonment of grazing results in the development of a closed sward unsuitable for Redshank. This is consistent with observations at other sites and described as State 4 (abandoned) saltmarsh (Section 4.4.4). On the other hand, the reintroduction of moderate to light grazing on ungrazed areas, or its continuation on others, results in a more structurally diverse sward. This can have beneficial effects on the redshank breeding population especially where cattle are used. Although they trample nests, the higher breeding densities the structural diversity affords more than make up for the losses (Norris et al. 1997).

Looking at the issue more widely, an increase in grazing pressure in Great Britain between 1985 and 1996 correlated with a decline in breeding Redshank. This was most marked on saltmarsh sites, where grazing patterns changed from lightly grazed to moderately/heavily grazed (Norris et al. 1998). Another study, in southern England, suggests that introduced Sika deer (*Cervus nippon*) can have an adverse impact on the suitability of saltmarsh for breeding Redshank. In this case, grazing significantly affected both the structure and species diversity of the sward. In ungrazed areas, *Spartina anglica* was dominant, helping to create suitable cover for nesting redshank (Hannaford et al. 2006). It seems likely that deer produce a tight low-growing sward similar to sheep-grazed areas, which is too open for redshank nests.

### 7.3.4 Grazing by Domestic Stock, Effects on Avian Herbivores

Under 'natural' conditions, saltmarsh use by grazing ducks and geese tends to be associated with the lower successional stages in vegetation development. Studies of changes in vegetation during saltmarsh succession in the Schiermonnikoog National Park, Wadden Sea, the Netherlands, showed that as sedimentation takes place, plant communities change from 'low marsh' species such as *Puccinellia maritima* into an *Atriplex portulacoides* marsh. At higher levels, the succession is from a *P. maritima* dominated sward to one with *Festuca rubra*. *Artemisia maritima* and eventually *Elytrigia atherica*, becomes dominant after about 60 years. A reduction in the use by Brent Geese, which preferentially graze the younger saltmarsh, occurs as succession takes place in the absence of grazing by domestic stock. Ultimately, the vegetation becomes unsuitable for grazing by the geese (Drent & van der Wal 1999).

By contrast, experimental saltmarsh grazed by sheep showed that captive geese preferred vegetation that was grazed in the spring, rather than ungrazed vegetation. Reductions in sheep grazing, and the accompanying increase in canopy height, led to a decrease in the levels of goose grazing, on both enclosed polder and saltmarsh. Plant communities intensively grazed by livestock and dominated by *Festuca rubra* and *Puccinellia maritima*, experienced higher grazing pressure by geese than long-term ungrazed or lightly grazed saltmarshes (van der Graaf et al. 2002).

A similar picture emerges from other studies in the Wadden Sea, where Dark-bellied Brent Geese and Barnacle Geese both preferred areas grazed by livestock, to ungrazed vegetation. This was true for both saltmarshes and agricultural fields (Loonen & Bos 2003). On the saltmarshes of the Hamburger Hallig in Schleswig Holstein, Germany, lower grazing pressure by Barnacle Geese occurred after the cessation of sheep grazing (Stock & Hofeditz 2003), an effect that was more pronounced in autumn than in spring. This general trend reflects the change in saltmarsh vegetation, which results in the eventual creation of a dense, matted turf unsuitable for avian herbivores, when a major reduction or cessation in grazing pressure occurs.

Shorelark, Snow Bunting (*Plectrophenax nivalis*) and Twite favour the lower saltmarsh vegetation and drift lines during the winter. They also visit intensively sheep-grazed upper saltmarshes, which resemble lower saltmarshes in their plant composition. However, despite this, the seed production per plant is much lower here, with some important seed producers hardly producing any seed at all. A combination of enclosure and increased grazing pressure is implicated in the reduction of the populations of these species since the 1960s. More recently, there is some evidence to suggest that the wintering populations of these species have increased due to a reduction in grazing pressure and cessation of enclosure (Dierschke & Bairlein 2004).

### 7.3.5 Grazing and Invertebrates

The invertebrate fauna is intimately associated with the different plants species as well as the tidal position of the saltmarsh. Grazing has important knock-on effects on the distribution of invertebrates across the saltmarsh. In particular, higher grazing pressures tend to shift the plant communities towards tillering grasses, including those that generally dominate the lower saltmarsh zones. The number of species present therefore diminishes in accordance with the lower number of invertebrate species associated with these zones (Table 11, Section 4.5.1). In

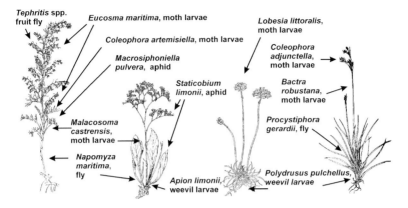

**Figure 53** A few of the phytophagous (plant-eating) invertebrates, feeding on selective parts of four 'higher' saltmarsh plants (after Dijkema 1984)

addition, as the saltmarsh is grazed, the upper parts of individual plants are removed and with them, their associated species (Figure 53). In Great Britain, species such as *Seriphidium maritimum* also have rare invertebrates such as the Essex Emerald (*Thetidia smaragdaria*) associated with it.

Change in the invertebrate fauna can also occur when a reduction in grazing pressure occurs. For example, in the saltmarshes of the Mont Saint-Michel Bay (France) an invasion of *Elymus athericus* took place over a 10-year period following a reduction in grazing pressure. The change in vegetation cover altered the density and flood resistance of the dominant halophilic spiders *Arctosa fulvolineata* (nocturnal lycosid) and *Pardosa purbeckensis* (diurnal lycosid). Comparisons in both uninvaded and invaded habitats showed a reduction in the population of *P. purbeckensis*, whereas there seemed to be a positive affect on *A. fulvolineata*. Vegetation structure, availability of prey and increased interspecific competition, all influenced these differences (Pétillon et al. 2005).

Removal of material in an experiment to mimic the effects of haying on saltmarshes in Massachusetts showed how surface invertebrates such as the amphipod Common Marsh Hopper (*Orchestia grillus*), leafhoppers (Order Homoptera) and the isopod *Philoscia vittata*, some of the most common invertebrates encountered in the marsh, decreased in abundance with the removal of saltmarsh vegetation (Ludlam et al. 2002). Losses of invertebrates, breeding birds and other animals sheltering in the more structurally diverse vegetation generally, increase as the grazing pressure becomes more intense.

## 7.3.6 Grazing and Sea Bass

There are few studies showing a direct effect of grazing on saltmarsh vegetation as it affects a marine animal. One exception is the Sea Bass (*Dicentrarchus labrax*)

(Laiffaille et al. 2001). In Mont Saint-Michel Bay, France, juveniles of the species forage inside the marsh on spring tides, feeding mainly on a detrivorous amphipod the Beachflea or Beach-hopper (*Orchestia gammarellus*) in the ungrazed marshes. In grazed areas, juvenile sea bass consume less food because of the replacement of this amphipod by other less suitable species (Laffaille et al. 2000).

## 7.4 Lesser Snow Geese (*Chen caerulescens caerulescens*) – A Conservation Dilemma?

This section provides a brief consideration of the impact of major increases in natural avian herbivores on the saltmarsh habitat, particularly from North America. In the USA, the Migratory Bird Treaty Act of 1918 set out to protect migratory species including many ducks and geese from hunting. Included within this list were all the main species of migratory geese including the Lesser Snow Goose, which breeds in the northern latitudes of Canada and the Arctic, migrating south for the winter (Figure 54). The Migratory Bird Conservation Act, and other laws such as the Duck Stamp Act 1934, (provides a mechanism for generating money for the acquisition and protection of important migratory bird habitats) helped establish further protection. The success of this legislation led to a large increase in the numbers of geese.

### 7.4.1 A Population Explosion

Changes in agricultural use, (in this case on the Gulf coast where this population overwinters), plus the protection afforded to the species at staging points on the migratory route, help to increase winter survival (Abraham et al. 2005; Jeffries et al. 2003). As with the European examples (Section 7.2.1 above), there is no natural population control during the winter. The increased use of fertilisers, the creation of palatable grass and cereal crops provide easily accessible and nutritious food. As a result, there has been a massive increase in the population over 30 years, tripling to nearly 6 million birds. The Lesser Snow Goose is one of several goose populations causing concern in North America. It is the most significant because of the high population, and the fact that it nests in coastal areas where it can cause complete loss of saltmarsh vegetation.

### 7.4.2 Grazing Impacts on Saltmarsh

Grazing by native herbivores, such as geese, is a natural occurrence and as indicated, some species preferentially graze saltmarshes, seeking out the more palatable grasses. It is a key herbivore in coastal ecosystems of Hudson and James Bays in Canada, where it grazes saltmarsh vegetation. This increase has led to

## 7.4 Lesser Snow Geese (*Chen caerulescens caerulescens*) – A Conservation Dilemma?

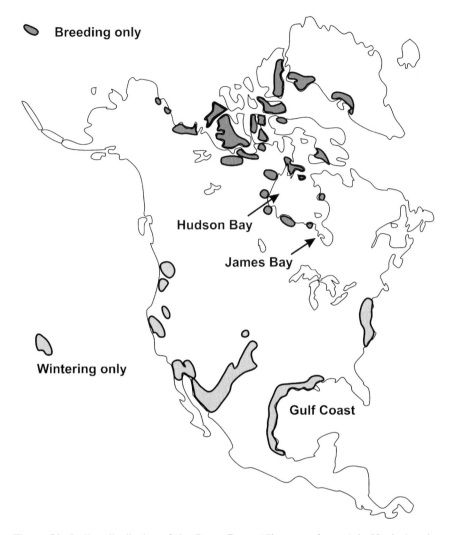

**Figure 54** Outline distribution of the Snow Goose (*Chen caerulescens*) in North America. Redrawn from the 'Snow Goose Range Map' on the Environment Canada, Wildspace (see http://www.on.ec.gc.ca/wildlife/wildspace/intro-e.html)

Arctic habitats, notably saltmarshes, being in danger of destruction because of goose grazing. Overgrazing and overgrubbing by geese cause changes in soil salinity and moisture levels that in extreme cases expose the underlying soil and cause an increase in soil salinity (Cargill & Jefferies 1984).

Severe environmental degradation of the affected Arctic landscapes alters plant community structure, succession and soil properties. Soil compaction, in particular, prohibits the natural recolonisation of native saltmarsh plants, even when bird numbers are reduced (Handa & Jefferies 2000). Large portions of the Arctic ecosystem have suffered possibly irreversible ecological degradation in this way. For example, at nine widely separated study sites, 35,000 ha of intertidal saltmarsh have been lost. Similar changes have occurred elsewhere along the 2,000 km coastline where the geese breed or rest during migration (Jefferies et al. 2006).

There is evidence to suggest that survival of goslings of the Lesser Snow Goose and adult females relies, in part, on the greater nutrient and fibre content of the saltmarsh vegetation, when compared to inland swards. Other evidence supports this suggestion. Overgrazing, which results in the destruction of saltmarsh, may be a factor in the observed weight loss of goslings. This in turn has implications for their survival during migration (Packham & Willis 1997).

### 7.4.3 Controlling Goose Numbers

The impacts on saltmarsh and other habitats were sufficiently significant for an Arctic Goose Habitat Working Group to be established. This recommended the reduction of the mid-continent white goose populations by 5–15% each year. It also recommended an extension of the harvest of snow geese by southern hunters beyond the restrictions in the Migratory Bird Treaty (US Fish & Wildlife Service). The full report 'Arctic Ecosystems in Peril' by the Arctic Goose Habitat Working Group is accessible, (see http://www.fws.gov/migratorybirds/issues/arcgoose/tblconts.html). There are also several web sites devoted to this issue, see for example:

- http://www.pnr-rpn.ec.gc.ca/nature/migratorybirds/dc00s04.en.html. This web address forms part of Environment Canada's Internet resource for weather and environmental information. It provides information and links on the impact of Lesser Snow Geese on arctic habitats;
- http://www.fws.gov/migratorybirds/issues/snowgse/tblcont.html. US Fish and Wildlife Service, Division of Migratory Bird Management has made available a Draft Environmental Impact Statement on Light Goose Management;
- http://research.amnh.org/~rfr/hbp/index.html. The Hudson Bay Project 'Ecosystem Studies and Conservation of Coastal Arctic Tundra' has technical and up-to-date information on the snow goose problems;
- See also http://icb.oxfordjournals.org/cgi/content/full/44/2/130 for an online copy of an up to date article by Jefferies et al. (2004).

## 7.5 Changing Biological Values

The above discussion shows that complex relationships exist between native and domestic grazing animals in relationship to the biological values attached to saltmarshes. The general conclusions are somewhat confusing, as grazing can cause both an increase and decrease in plant diversity, for example:

- Heavy grazing may eliminate sensitive species, producing a dense cover of grasses, especially in higher saltmarsh;
- In some marshes, trampling produces bare patches allowing annuals and other low marsh species to invade;
- Plant succession can be retarded because of heavy grazing;
- Intermediate levels of grazing by sheep, cattle and horses tend to produce communities with a high diversity of species richness derived from both spatial and temporal diversity;
- Grazing by geese creates bare areas with higher salinity soils, only suitable for salt-tolerant species. Their removal can result in an increase in species richness in sub-arctic saltmarshes;
- Invertebrate herbivores may also have a greater influence than generally supposed.

This has further effects on other elements within the saltmarsh, particularly in relation to the numbers and diversity of animals living in or on the vegetation. These relationships suggest that the management and restoration of saltmarsh is equally complex and promotion of one interest can have adverse consequences for another.

In order to help unravel some of these complexities, a simply Grazing State Evaluation Model is proposed. This uses the 'states' of vegetative saltmarsh under different grazing regimes (Section 4.4), which can be summarised as follows:

- State 1 – heavily grazed (Figure 21), characterised by short swards, compact vegetation, lack of structural diversity;
- State 2 – moderately grazed (Figure 22), include mosaics of vegetation with some more heavily grazed and others less heavily grazed swards;
- State 3 – historically Ungrazed / Lightly grazed (Figure 23) good structural and plant species diversity;
- State 4 – abandoned, formerly grazed (Figure 24) increasing growth of coarse grasses, leading to dense vegetation and reduction in plant species diversity;
- State 5 – overgrazed, is an extreme form of State 1 saltmarshes. Overgrazing can cause 'eat-outs' with a complete loss of the vegetation, hypersaline conditions and in some cases erosion.

A similar approach, adopted in relation to land use management in the Wadden Sea, defined different grazing pressures in relation to the height of the vegetation and its heterogeneity. Thus, 'intensive grazing' equated to overall short swards, 'no grazing' to an overall tall canopy and 'moderate grazing' with a pattern of low sward and tall canopy. These definitions represent a truer reflection of the response of the vegetation to grazing, than using some predefined stock level (Bakker et al. 2005).

These conditions impact on other nature conservation attributes notably in relation to winter grazing by ducks and geese, nesting birds and invertebrates, including spiders as detailed above.

## 7.6 'Grazing' State Evaluation Model

The nature of the vegetation forms the basis for this evaluation model. In this model, grazing is the key driver used to distinguish the states. In particular, the State Evaluation Model is concerned with the way in which grazing alters the structure, vegetation type, numbers and diversity of plants and animals. Understanding this is an essential prerequisite for decisions on altering stock use and/or undertaking restoration in saltmarshes with degraded vegetation. The key areas of interest for each state are summarised in Figure 55. The columns indicate the level of nature conservation interest ranging from little or no interest 0, on the $y$-axis to 3, significant interest. These values relate to the first four states only, described above.

Lightly grazed saltmarshes or those historically ungrazed tend to have the greatest structural diversity and most diverse fauna and flora. At the other end of the spectrum, heavily grazed saltmarshes are structurally simple and with the exception of some upper and transitional communities, tend to be impoverished. Heavy grazing may provide enhanced conditions for herbivores, especially ducks and geese. However, any rapid cessation (abandonment) of grazing leads to further impoverishment as the vegetation responds to the relaxation in grazing pressure. The growth of the more aggressive species may result in the vegetation becoming rank and unsuitable for grazing, as well as lacking most of the species associated with less heavily grazed sites. These conditions provide the basis for developing a second, vegetation State Evaluation Model.

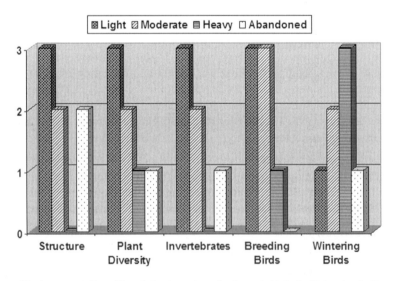

**Figure 55** A representation of the relative importance for the main biological interests of saltmarshes, under different grazing regimes. **1** denotes limited, **2** moderate and **3** high biodiversity. Note the numbers 1–3 are indicative only and based on the author's experience of these habitat types

## 7.6 'Grazing' State Evaluation Model

### 7.6.1 Effecting Vegetative Change – A State Evaluation Model

In order to determine the appropriate management option, it is important to assess the desired state in relation to the principal values affected by the move from the existing state. To aid this process, the first four states identified summarise the principal pathways for change and movement between the states (Figure 56).

The principal difference is between heavily grazed saltmarshes and lightly grazed ones. These have fundamentally different values as detailed above. In this context, a significant question arises as to whether to graze a saltmarsh or not. Managing and/or restoring former grazing regimes in accordance with this model forms the basis for the discussion in Chapter 8.

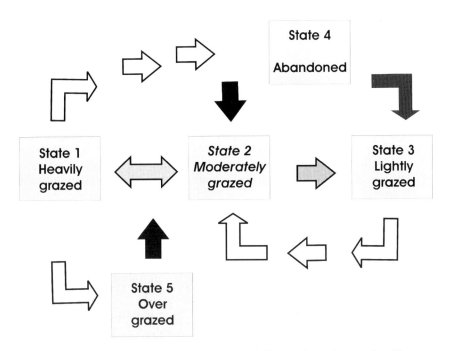

**Figure 56** Summary of the pathways for change, in relation to saltmarsh vegetation. The arrows indicate the principal routes for change: open arrows imply loss of interest; black arrows the route to restoring this interest; grey arrows suggest movement between states having different attributes but not necessarily representing major adverse change

# Chapter 8
# Grazing Management

## To Graze or not to Graze

## 8.1 Introduction

Grazing by both native herbivores and domestic stock is an important element in the development of saltmarshes in temperate regions. By removing the vegetation, the animals significantly alter the value of the habitat. Changes in grazing pressure can have both dramatic and subtle affects on the nature of the vegetation and its associated animals. As such, setting or changing grazing levels on a particular saltmarsh requires careful planning.

A key question when considering management or restoration of saltmarsh is, has the vegetation been grazed or not? Historically, ungrazed (by domestic animals) or lightly stocked saltmarshes tend to have the richest flora and fauna (Sections 4.5). These habitats are restricted to locations where human occupation is limited or on inaccessible islands with fringing saltmarshes. In Europe, grazing by domestic stock extends across the UK, especially in the west and into northern Scotland, Ireland, around the Baltic Sea (Dijkema 1990) and in the Wadden Sea. It is also prevalent around most estuarine saltmarshes in the rest of Europe (Dijkema 1984) though most intensive in northern and central areas (Beeftink 1977b).

The 'grazing' State Evaluation Model suggests there are four principal reasons for wishing to manipulate grazing on a saltmarsh:

1. The saltmarsh has an existing grazing regime, which could be changed to enhance elements of nature conservation interest;
2. A recent increase in grazing on a lightly/ungrazed saltmarsh has led to a reduction in biological diversity;
3. A cessation of grazing results in structural change and a loss of species associated with a heavily or moderately grazed saltmarsh;
4. Vegetation is destroyed because of overgrazing.

This chapter deals in some detail with the assessment of grazing in relation to maintaining existing saltmarsh values in the face of external forces driving change. It also seeks to provide information on the methods, which are appropriate when considering the restoration of saltmarsh vegetation, including the reintroduction of grazing on abandoned sites.

## 8.2 Assessing the Need for Change

Understanding the historical use of the site is the first stage in any assessment of the need for intervention. Normally, the reference point will be the stated nature conservation value of the site, either in its own right or as part of a wider coastal system. Assuming that grazing management is appropriate to the nature conservation requirements of the site, no intervention is required, other than to ensure maintenance of the existing regime. The need for changing grazing management on an individual saltmarsh begins with the identification that an adverse change has taken place, affecting the nature conservation value of the site. This may be the result of gradual change in farming practices bringing about a reduction, cessation or increase in stock densities. Surveillance is likely to provide the first indication that adverse change is occurring. Thereafter monitoring, including academic research, provides the means of assessing the impact of the changes and the effectiveness of intervention to reverse them.

### 8.2.1 Historical Understanding

Grazing by 'native' herbivores will almost certainly have taken place on most saltmarshes from their early development. It seems likely that these would have been subtle changes affecting the balance between different species, as suggested by the studies of Brown hares and geese (Section 7.2.2), rabbits and rare plants (Section 7.3.2), deer and Redshank (Section 7.3.3). We know that in the latter half of the twentieth century goose numbers have had an adverse and major impact in North America (Section 7.4). Even in prehistoric times, Pleistocene herbivores probably reduced saltmarsh biodiversity, in a way similar to present day animals, such as horses (Levin et al. 2002).

In the absence of grazing by domestic stock, historically ungrazed or lightly grazed (State 3 saltmarshes), tend to have the greatest biological diversity (Section 7.3). The so called 'natural landscapes' appear to have developed in response to geomorphological conditions not affected by human activity (Bakker et al. 2005). A similar situation occurs in lightly or ungrazed saltmarshes in North Norfolk, south-east England (Figure 57).

Conversely, heavily grazed (State 1 saltmarshes) have a restricted flora dominated by a few grazing tolerant species, an impoverished fauna, but with the possibility of an enhanced population of avian herbivores. Thus, in assessing the need for intervention in grazing management, the historical patterns of grazing and their influence on plant succession, are required. It may not always be possible, due to the difficulties of inferring successional pathways from vegetation mosaics (Adam 1990), to determine the precise changes that have taken place. However, the general state of the vegetation (Section 4.4) provides an indication of the most relevant features. Grazing changes these biological values, which are summarised in Section 7.5.

**Figure 57** Species-rich saltmarsh with a variety of herbs such as *Limonium vulgaris*, *Triglochin maritima*, *Atriplex portulacoides*, a rich invertebrate fauna and breeding birds such as Skylark

## 8.2.2 Protecting Nature Conservation Values

In Europe, the presence of grazing can be an important factor in the selection of sites for designation as nature conservation areas. For example, the description of the Wash and North Norfolk Coast Special Area of Conservation (south-east England) designated under the European Union Habitats Directive, includes the following statement:

'This site on the east coast of England is selected both for the extensive ungrazed saltmarshes of the North Norfolk Coast and for the contrasting, traditionally grazed saltmarshes around the Wash' (Joint Nature Conservation Committee web site, see http://www.jncc.gov.uk/page-1458 – EU Code UK0017075). Listed among the operations requiring consent from Natural England (English Nature), the 'Damaging Operations' considered likely to adversely affect the interest of the Wash Site of Special Scientific Interest (SSSI), are:

> "The introduction of grazing and changes in the grazing regime (including type of stock or intensity or seasonal pattern of grazing and cessation of grazing)'. English Nature web site (see http://www.english-nature.org.uk/Special/sssi/), SSSI identification code, 1002591."

The same condition applies to the Morecambe Bay SSSI and Special Area of Conservation, which:

> "is characteristic of saltmarshes in northwest England, with large areas of closely grazed upper marsh'. English Nature web site (see http://www.english-nature.org.uk/Special/sssi/), SSSI identification code, 1001807."

Both sites have significant numbers of grazing geese. In the Wash, these include Pink-footed Goose, 14.8% of the Eastern Greenland/Iceland/UK population and Brent Goose 7.4% of the Western Siberia/Western Europe population. In Morecambe Bay there are large numbers of wintering waterfowl, including Pink-footed Goose, with an average of 2,475 individuals, 1.1% of the Eastern Greenland/Iceland/UK population.

In the Wadden Sea (southern North Sea), many of the bird species exceed the 1% waterfowl population criterion used in the designation of internationally important sites and include the avian herbivores such as the Brent Goose and Wigeon.

### *8.2.3 Assessing the Implications of Grazing*

Monitoring studies on the grazing effects of native and domestic animals are reported from the Wadden Sea, sites in England, notably the Wash and in the Bay St Michele (France). Interpreting their significance for nature conservation management is not always easy. A study of intensively sheep-grazed, mainland saltmarshes in Schleswig-Holstein (northern Germany) showed how the vegetation responded to the introduction of different intensities of grazing over a four-year period (Table 17).

From this study, the authors reach the conclusion that 'in terms of nature conservation cessation of grazing is recommended' (Kiehl et al. 1996). Four years is too short a timescale for such a recommendation. Studies undertaken over longer periods, including those referred to in Section 7.3.1, come to rather different conclusions. Research in France showed that after 6 years *Elytrigia atherica*, though present and not dominant, was likely to become so in a few years (Tessier et al. 2003). After 20 years, in the mainland saltmarshes of the Wadden Sea, the same community can become dominant as the number of plant communities decreases (Bakker et al. 2003). Given that the dominance of *Elytrigia atherica* can lead to an impoverishment in the flora and fauna on saltmarshes where grazing has

**Table 17** Summary of the changes in vegetation from a series of permanent plots recorded from 1989–1993 (Kiehl et al. 1996)

| Grazing intensity | Number of sheep | Vegetation |
| --- | --- | --- |
| Intensive | 10 per ha | Short *Puccinellia maritima* dominated sward with *Salicornia europaea* and *Suaeda maritima*. *Atriplex portulacoides* and *Aster tripolium* rare. |
| Moderate | 3.0 per ha | *Puccinellia maritima* remains dominant Density of *Salicornia europaea* reduced, *Suaeda maritima* optimal growing conditions. *Atriplex portulacoides* and flowering *Aster tripolium* both cover and size increases towards the sea |
| Light | 1.5 per ha | Increase in vegetation height towards the sea due to reduction in grazing at lower tidal levels |
| Cessation | No grazing | *Puccinellia maritima* is successively replaced by *Festuca rubra*, *Atriplex portulacoides* and *Aster tripolium*; greater structural diversity |

ceased (Section 7.3.1), the recommendation to cease grazing, based on only four years data, may appear surprising.

These and other similar studies, highlight the fact that it is difficult to predict longer term changes in vegetation from short-term studies. In this case, the original conclusion to cease grazing did not improve nature conservation values in the longer term. The starting point is also important. Grazing has probably taken place, in summer at least, on the mainland saltmarshes of the Wadden Sea as far back as 600 BC. As has already been described, this changes the vegetation in quite profound ways, such that any relaxation in grazing pressure may have different outcomes, depending on the dominant species present at the time the change occurred.

The extensive studies undertaken in the Wadden Sea and elsewhere throw some light on this, as described in Chapter 7. They are also discussed in a paper 'To graze or not to graze: that is the question' (Bakker et al. 2003). This provides detailed descriptions of the way the vegetation changes in response to different grazing pressures. However, they conclude that it is only possible to identify the appropriate grazing regimes once the targets for management are established. They suggest that the nature conservation question is not: 'to graze or not to graze', but 'what are the targets and which role can grazing management play to reach these targets?'

## 8.2.4 Grazing Management in North America

The significance of grazing management to the development of saltmarsh vegetation in Europe is well established. There is much less information on the influence domestic stock in North America has on vegetation development. Descriptions of the way of life of the Indians when the settlers arrived from Europe in the 1500s, suggest that geese were part of the diet. In his *Key into the Language of America*, published in England in 1643, Roger Williams describing the Narragansett word for geese as 'Honck, honckock', observed, 'that the Indians hunted a variety of waterfowl, including swans, brants and ducks'. Given the wooded nature of the interior, they were probably associated with the open saltmarsh.

Haymaking certainly took place historically (Section 2.1.4). Salt hay farms occurred in Delaware Bay, on areas of saltmarsh lying at the highest tidal levels. They occurred as flat, firm turf sometimes with dikes around them to restrict flooding (Teal & Weinstein 2002). There is no mention, as to whether grazing by domestic stock, also took place.

In Narragansett Bay, there is reference to the historical use of 'Walker Farm Marsh' for grazing livestock, according to the 'archive of restoration projects' (see http://www.estuaries.org/). Saltmarshes near Galveston, Texas, include studies of the effects of cattle grazing on historically grazed saltmarsh vegetation and associated animals (Martin 2003). A search of the literature has so far found no further mention of grazing, as a major factor in saltmarsh management in North America.

## 8.3 Managing Grazing Levels

The Grazing State Evaluation Model suggests there is interplay between different levels of grazing such that the saltmarsh values change in response. This is certainly true, although for most sites there is an existing pattern of grazing linked to the identified nature conservation values. This will form the basis for management. Deviation from these established values will form the basis for decisions to intervene and change the grazing management regime on the saltmarsh.

For many sites, the prevention of damage from development will be the focus for conservation action. Thereafter, especially in nature reserves and statutorily designated sites, maintenance of a grazing regime to retain the identified interest, will form the basis for management. Where adverse changes become evident then intervention to reverse them will be required.

The following section provides information to help identify the most appropriate form of grazing management. Please note that the figures provided below are indicative only. The precise grazing regimes will depend on the size of the saltmarsh and its physical condition, including the presence of creeks. These, together with tidal range, will affect the extent and period of access and hence grazing pressure in different locations. The location of watering points also influences the intensity of grazing nearby.

### 8.3.1 Maintaining Ungrazed/Lightly Grazed 'Natural' Saltmarsh (State 3)

Lightly grazed saltmarsh is probably nearest to the natural condition. There will normally be no history of grazing by domestic stock. Grazing is limited to native animals such as Brown Hares in Europe, ducks and geese in Europe and North America and Muskrat (*Ondatra zibethicus*) in North America.

Historically, ungrazed saltmarshes are limited in both number and size. Grazing by domestic stock has probably never taken place on some island saltmarshes of the Wadden Sea (Dijkema 1984). Although the exception rather than the rule, these can be some of the most visually distinct and species-rich saltmarshes (Section 4.4.3; Figure 57). Other examples include the smaller high-level saltmarshes associated with beaches on the south-west coast of Portugal and in the Mediterranean (Figure 8). There is no direct information on past grazing practice, but there seems to be little historical demand for grazing on saltmarshes. Upland pastures provide the summer grazing through the operation of the traditional practice of 'transhumance', a system involving the movement of domestic animals away from the extremes of climate. In 'ascending transhumance', animals spend the colder months near the coast moving to the uplands during the hot dry summer months. 'Descending transhumance' works in the opposite direction with animals leaving the mountains for the lower pasture during the colder winter months (Blondel & Aronson 1999, Box 8.2, page 218).

The appropriate stocking density for lightly grazed saltmarshes is:

> "no more than 2.0 sheep or 0.3 cattle per ha grazing for 6 months of the year' (Beeftink 1977a)."

It is likely that for historically, ungrazed saltmarshes even these levels may have an adverse impact on some species.

### 8.3.2 Maintaining Moderately Grazed Saltmarsh (State 2)

Managing saltmarsh that has a history of grazing by domestic stock is largely one of manipulating the grazing regime. In many locations, moderately grazed saltmarsh has a range of biological diversity. They may or may not have some of the rarer plant species associated with historically ungrazed or lightly grazed areas. However, the presence of a greater range of habitats for invertebrates, breeding birds and even grazing ducks and geese in the more open and accessible parts of the site, provides compensation. The range of stocking densities associated with this type of grazing is:

- 5–6 sheep, 1.0–1.5 young cattle per ha from April – October (Beeftink 1977a) or
- 0.6 cattle year round (Kleyer et al. 2003)

In addition to setting grazing levels on saltmarshes, where creeks provide 'wet fences', differential grazing patterns emerge. Saltmarsh management can achieve the same 'moderate' grazing regime by differentiating the saltmarsh into ungrazed and lightly grazed areas, each representing 50% of the habitat. This is equated with an overall grazing pressure of less than 0.5 cattle per ha on larger saltmarshes (Andresen et al. 1990).

State 2 saltmarshes provide a range of opportunities for a wide variety of plant and animal species, giving good representation of the values described in Section 4.5.1.

## 8.3.3 *Maintaining Heavily Grazed Saltmarsh (State 1)*

Grazing levels of up to 6.5 sheep per ha year round, 9–10 sheep or 2 cows per ha in summer (Section 4.4.1) lie at the upper end of the spectrum of stocking regimes. They approach those of inland intensively grazed grassland. As with lightly grazed saltmarsh, if this is the traditional use of the saltmarsh and the site has important populations of avian herbivores, then there should be a presumption of maintaining the State 1 regime.

## 8.4 Modifying Saltmarsh Vegetation

Changes to the stock regimes (mainly of sheep and cattle), will cause a shift in nature conservation value as summarised in the 'grazing' State Evaluation Model (Figures 55 and 56). The trends and trades-offs, discussed in Chapter 7 are complex. For example, increases in grazing pressure tend to affect the early and middle stages of succession by helping to extend the dominance of grass species over herbs. The elimination of grazing-sensitive species more generally from the sward occurs as livestock numbers increase, further diminishing the diversity of plant species in the mid- to upper-communities. Conversely, some species may benefit from a more open sward as trampling and other factors create gaps in the canopy as illustrated in Figure 29 above. An increase in grazing pressure can also cause a loss of suitable breeding habitat for Redshank, as structural diversity is decreased (Norris et al. 1998). Conversely, for saltmarshes with a dense monoculture of species such as *Elytrigia atherica*, an increase in stock levels may improve suitability for breeding Redshank.

The balance between the 'natural' values associated with heavily grazed and more moderately grazed saltmarshes are important for nature conservation values. The issues surrounding any movement between State 1, 2 or 3 saltmarshes will involve changes in stock density, type of animal and timing of use. The breed of cattle or sheep (the main domestic grazing animal) may also have an influence on the biological diversity. The following discussion covers some of the more important situations, when making decisions to modify an existing grazing regime might be desirable.

## 8.4.1 *Historically Ungrazed or Lightly Grazed (State 3) and Moderately Grazed (State 2) Saltmarsh*

Where saltmarshes, which have historically been lightly or ungrazed, it is unlikely that any reversal of the State 3 – State 2 would be desirable. Indeed, the introduction of grazing by domestic stock will almost certainly reduce their value. In addition to the loss of the upper parts of the plants, and with it the species associated with them, grazing-sensitive species such as *Limonium vulgare* or *Atriplex portulacoides* may be eliminated from the vegetation altogether as grazing pressures increase (Section 7.3.1).

The 'grazing' State Evaluation Model suggests that intervention could be required when there is a change from historically State 3, lightly grazed to moderately grazed saltmarsh, which would tend to reduce the vegetation structure and overall biodiversity. The general trend in recent years is away from grazing livestock on marginal land, such as saltmarshes. Thus, the need to reverse such a change rarely happens. If such a situation does occur, then a return to the historical levels of grazing, or no grazing at all, would appear to be the most prudent course of action.

## 8.4.2 Reducing Grazing Pressure – Heavily Grazed (State 1) Saltmarsh

Saltmarshes heavily grazed by sheep have an impoverished low- to mid-level vegetation. At the highest stock densities, grazing and dunging results in a tight low sward with no structural diversity (Figure 34). Subject to the caveat in Section 8.3.3 concerning the maintenance of stock levels on sites with significant populations of avian herbivores, reducing the grazing pressure is the main option. The results of such an incremental reduction include an increase in the overall height of vegetation and improvement in its structure. This in turn leads to an increase in the diversity of plant and animal species, typically associated with State 2 saltmarshes.

There are no established methodologies for setting grazing regimes. The key to modifying the existing regime of a State 1 saltmarsh lies in making decisions about the desired outcome in relation to the type of nature conservation interest to be encouraged. Looking at the way, grazing changes the balance between the various elements within the saltmarsh (Chapter 7) provides an indication of the options and opportunities. Changing the grazing levels should be undertaken gradually and the effects monitored. This must take into account seasonal variations in climate because of the influence of water logging.

Evidence from the Wadden Sea provides an indication of the way changes occur. Here, following the establishment of a nature reserve, the maintenance of drainage ditches and a reduction in grazing pressure took place, in an attempt to restore natural saltmarsh vegetation on a 'man-made' saltmarsh (Section 2.4.3) in the Dollard estuary, the Netherlands. Monitoring the effects 8–9 years after the change in management showed the effects on the vegetation of the distribution of cattle across the saltmarsh and the impact of water logging. Cattle grazed most intensively close to the sea wall. Whilst *Elytrigia atherica* (*Elymus atherica*) decreased, *Aster tripolium* increased at these upper levels. At the same time, the combination of a moderate stocking regime and neglect of the drainage system helped create bare patches and conditions favouring earlier successional stages in saltmarsh development. This regime also prevented the dominance of *Elytrigia atherica* in the brackish sections of the marsh. To seaward *Spartina anglica* and Sea Club-rush (*Bolboschoenus maritimus*) declined, partly due to the expansion of *Phragmites australis* and because of grazing by Greylag Geese (*Anser anser*) (Esselink et al. 2000, 2001). These changes benefited birds breeding in reed beds, as well as halophytic plants, breeding Redshank and grazing waterfowl, because of the mosaic of habitats created.

The least beneficial change occurs with the abandonment of grazing. This has adverse consequences as discussed in Section 7.3 particularly for grazing ducks and geese. Although there is some apparent improvement in plant and animal species diversity in the first few years, many of the values are lost in a relatively short period of 20 years (Section 7.3.1.). Reversing the resulting growth of matted coarse vegetation is one of the more frequently encountered management activities, dealt with next.

## 8.4.3   *Restoring Grazing on Abandoned (State 4) Saltmarshes*

The trend, in Europe at least, is for the use of marginal agricultural land, including saltmarsh, for intensive grazing to decrease. Abandoned grazing is the most frequently encountered situation requiring intervention, for nature conservation purposes. For example, up until the last 20 years or so, cattle and sheep grazed the majority of the mainland saltmarshes in the Wadden Sea. Since then grazing has ceased on large areas, due to lack of interest from farmers (Dijkema & Wolf 1983).

Changes in the emphasis away from agricultural use, to the establishment of nature conservation areas, also resulted in the abandonment of grazing in some areas (Bakker et al. 2003). The papers discussed in Section 7.3.1 show that such a reduction in grazing pressure can lead to the growth of coarse grasses, making them suboptimal for wintering avian herbivores (Section 7.4.5). Once the grazing pressure is removed these tend to respond with vigorous growth leading to the matted dense turf prevalent in these areas (Figure 37). Since species preferentially graze the more nutritional tillering grasses, any reduction in grazing levels that result in coarsening of the vegetation could be detrimental to this interest, and result in the displacement of birds onto agricultural land.

Although there is an initial 'improvement' in the structure of the vegetation in some areas and the diversity of the invertebrate fauna increased, see for example in the Wadden Sea (Andresen et al. 1990), this may be a short-lived effect. A similar study of sheep grazing in the northern part of the German Wadden Sea showed similar short-term changes (Kiehl et al. 1996).

Certainly 'abandoned', formerly grazed saltmarsh becomes less suitable for grazing ducks and geese, often the main reason for their nature conservation designation. This represents a major trend in saltmarsh management leading to a requirement to restore the vegetation through the reintroduction of grazing or mowing (Section 8.4.4; Table 18).

**Table 18** Dominant plant groups in transects of 10 ha each from west to east in the Netherlands, Wadden Sea mainland saltmarshes

| Transects | 1960–1970 | 1970–1980 | 1980–1985 | 1985–1990 | 1990–1995 | 1995–2000 | 2000–2005 |
|---|---|---|---|---|---|---|---|
| FRIESLAND | Heavily grazed | | | | | Less grazing | Stopping drainage |
| 005–008 | | Pioneer | Climax | Low | Climax | Climax | Climax |
| 021–024 | | Pioneer | Low | Low | Climax | Diverse | Climax |
| 041–044 | | Pioneer | Pioneer | Pioneer | Pioneer | Diverse | Diverse |
| 053–056 | Pioneer | Pioneer | Low | Low | Low | Low | Diverse |
| 069–072 | Pioneer | Low | Low | Low | Low | Pioneer | Diverse |
| 085–088 | Pioneer | Low | Low | Low | Low | Low | Diverse |
| 101–104 | Low | Low | Low | Low | Low | Low | Middle |
| 121–124 | Low | Low | Low | Low | Low | Low | Low |
| 145–148 | Low | Low | Low | Low | Low | Low | Diverse |
| 167–170 | Low | Low | Low | Low | Low | Climax | Diverse |
| 205–208 | Pioneer | Pioneer | Low | Low | Climax | Climax | Climax |
| GRONINGEN | Light/moderate/ heavy grazing | | Less grazing | | | | Stopping drainage |
| 260–263 | Pioneer | Low | Low | Low | Pioneer | Pioneer | Pioneer |
| 286–289 | Low | Low | Diverse | Diverse | Diverse | Diverse | Climax |
| 308–311 | Low | Low | Diverse | Diverse | Diverse | Climax | Climax |
| 324–327 | Low | Low | Low | Low | Diverse | Diverse | Climax |
| 336–338 | Low | Low | Low | Low | Low | Diverse | Climax |
| 356–359 | Pioneer | Low | Pioneer | Pioneer | Low | Diverse | Climax |
| 372–375 | Low | Low | Low | Diverse | Diverse | Diverse | Diverse |
| 392–395 | Low | Low | Low | Low | Low | Climax | Climax |
| 412–415 | Low | Low | Low | Middle | Middle | Climax | Climax |
| 428–431 | Low | Low | Pioneer | Pioneer | Low | Low | Low |
| 448–451 | Pioneer | Low | Pioneer | Pioneer | Pioneer | Pioneer | Pioneer |
| 468–471 | Low | Low | Low | Diverse | Diverse | Climax | Climax |
| 488–491 | Low | Low | Low | Low | Low | Diverse | Climax |

Key: Pioneer plants with *Salicornia* spp. and *Spartina* spp; Low saltmarsh plants with *Puccinellia maritima* and *Atriplex portulacoides*; Diverse zones with *Aster tripolium, Spergularia* spp, *Triglochin maritimum, Limonium, Plantago* (= Asteretea); Climax plants *Elytrigia atherica, Atriplex prostrata*; Middle saltmarsh plants *Artemisia* (*Seriphidium maritimum*), *Armeria maritima, Juncus gerardii, Festuca rubra, Agrostis stolonifera, Glaux maritima*. The table shows how the vegetation succession develops from a pioneer, through a more diverse to a climax community, which when grazing and drainage stops eventually become dominated by *Elytrigia atherica* and *Atriplex prostrata* on very clayey soil (unpublished Table, Dijkema personal communication).

These situations raise a number of questions:

- Is the reduction in grazing affecting the value of a particular saltmarsh for winter avian herbivores?

- What is the significance for areas designated under national or international conventions or directives?
- What are the implications of any change to the management for the existing vegetation communities?
- Will it affect the range and variation of other species such as invertebrates?

The reintroduction of grazing is the preferred restoration option. The grazing regime adopted will determine the nature of the restored biodiversity interest. Depending on the desired outcome, it might be appropriate to move from State directly from 4 to State 1. In this situation, the approach will involve reintroduction of the more intensive level of grazing (Section 8.3.3). Restoring vegetation to a more open state can be difficult. Depending on the length of time from abandonment, the sward may be very dense. In these situations, high initial stocking levels will help open up the matted turf. A high groundwater table helps ameliorate these conditions. Following the introduction of grazing the reversal of the effects of abandonment include, the growth of smaller more palatable halophytes, such as *Puccinellia maritima* (Kleyer et al. 2003).

Intermediate, 'moderately' grazed saltmarshes that combine elements of State 1 'heavily' grazed and State 3 'lightly' grazed might have the greatest nature conservation benefit. This is referred to as 'mosaic grazing' and is being introduced as part of the restoration programme for the saltmarshes in Groningen where the climax vegetation includes unpalatable 'woody' plants (Table 18). A prudent restoration regime, therefore, might include introducing grazing levels appropriate to State 2 moderately grazed saltmarsh. Low-level, open-range grazing regimes, identified above as being appropriate to moderately grazed (State 2) saltmarsh vegetation, will help create the more diverse conservation interest. The grazing levels appropriate to this will depend on the desired outcome, but will range from 5–6 sheep, 1.0–1.5 young cattle per ha from April to October or 0.6 cattle year round as referred to in Section 8.3.2. At these moderate levels, the saltmarsh has the best chance of supporting a wide range of species with reasonable structural diversity.

### 8.4.4 Mowing as a Management Tool on Abandoned Saltmarshes

Both grazing and mowing enhance species diversity on abandoned saltmarsh (Bakker 1985). Mowing alone can reverse the growth of matted vegetation on abandoned saltmarshes. However, although mowing increases plant species diversity initially, after only 5 years, the plant diversity decreases again as *Festuca rubra* dominates the sward. In contrast, reintroduction of cattle grazing alone enhances species diversity through the gradual reduction in litter (Cadwalladr & Morley 1971; Cadwalladr et al. 1972).

As indicated above, mowing can achieve some of the desired conditions, but may not be sustainable over the long term. It is highly labour-intensive, depends on there being a level surface free from drainage creeks, and hence is only feasible over a relatively small area. On higher-level 'clayey' saltmarsh zones in the Netherlands, it is possible to maintain grasslands rich in freshwater species by mowing (Dijkema pers.com.).

## 8.4.5 Restoring 'Overgrazed' Saltmarsh

The situation in North America provides a different set of challenges. Here, the effect of 'eat-outs' by geese results in change in the saltmarsh habitat which makes it difficult for restoration to take place. A combination of complete loss of vegetation (including the root system), compaction of the soil and hypersaline conditions makes reintroducing plants difficult. In one study, removal of goose grazing resulted, in 5 years, in an increase in higher plant species from 6 to 16 species in a 5×5 m exclosure. This accompanied a rapid change in structure and litter accumulation (Bazely & Jefferies 1986). However, studies that are more recent suggest the problem is much more acute. It is argued that, without a reduction in the population of grazing geese, the saltmarshes will become irreparably damaged (Jefferies et al. 2006; Section 7.4.3).

## 8.5 Conclusions

The key determinant, when deciding on changing an existing grazing management regime, lies in assessing the historical and current patterns in relation to the nature conservation values. Decisions will depend on the extent to which adverse change causes loss of one or more attributes, as detailed above. Whatever the decision, it will be important when undertaking such a change to ensure that grazing is maintained at a level suitable to prevent the rapid growth of grasses (such as *Puccinellia maritima*, *Elytrigia atherica* and *Festuca rubra*).

In areas that provide palatable herbage for grazing ducks and geese, reducing grazing pressure may affect the carrying capacity of the marsh. In this context, it is important to recognise that a change in conservation status can take place very rapidly. A marsh seemingly supporting a varied flora and fauna can show a loss of interest in a matter of only a few years. A common saltmarsh grass *Puccinellia maritima* dominated sward, for example, which had persisted for at least 50 years, reverted to a dense *Elytrigia atherica/Festuca rubra* sward (State 4 abandoned) in only 10 years following cessation of grazing (Ranwell 1964).

The principal reasons for wishing to change a grazing regime in relation to maintaining nature conservation values lie in the condition of the vegetation and associated animals. In particular:

> "Grazing of previously ungrazed marshes can lead to loss of plant diversity.
> Overgrazing of grazed marshes can reduce food source for birds.
> No grazing can result in the dominance of competitive plant species, with the loss of other plants like Long-stalked Orache (*Atriplex longipes*).
> Tall growth of saltmarsh can reduce the suitability of feeding/roosting areas."

Taken from the Morecambe Bay European Marine Site, Scheme of Management (see http://morecambebay.com/), north-west England.

# Chapter 9
# *Spartina*

# Friend or Foe?

## 9.1 Introduction

*Spartina* species form a group of salt-tolerant grasses which occur in several different parts of the world. Geographically centred along the east coast of North and South America, outliers occur on the west coast of North America, Europe and Tristan da Cunha (Table 19).

**Table 19** Five of the fourteen or so **native** species of Cord-grass found in saltmarshes, their main areas of geographical distribution and tidal limits; MSL – mean sea level; MHW – mean high water mark

| English name | Latin name | Natural distribution |
|---|---|---|
| Smooth Cord-grass | *S. alterniflora* | Eastern USA Maine – Texas from 0.7 m below MSL to c. MHW |
| Salt Meadow Cord-grass | *S. patens* | Eastern USA from MHW to c. 0.5 m above MHW |
| California Cord-grass | *S. foliosa* | Bodega Bay – Baja, California, USA |
| Dense-flowered Cord-grass | *S. densiflora* | Chile, South America near MHW, or just below it on open mud |
| Small Cord-grass | *S. maritima* | British Isles, Europe up to MHW spring tides |

These species of grasses, have the ability to grow quickly and reproduce from seed, rhizomes or broken vegetative shoots. Once established they can colonise estuarine sand and mudflats quickly because of their ability to exploit areas outside the normal tidal range of the majority of perennial saltmarsh plants (Thompson 1991). The primary aim of this chapter is to consider the origins of the species, the reasons for its success and need for management.

### 9.1.1 *The Nature of Colonisation*

Most of the species (except the sterile hybrid *Spartina townsendii*) reproduce from seed. Plants flower quite late in the season, producing seed in the autumn. Colonisation by native species depends on the initial establishment of seedlings, which thereafter expand by vegetative means to form clones. *S. alterniflora* is one of the first colonisers of mudflats on the east coast of North America. In its native range, interspecific competition appears to limit its competitive ability at higher saltmarsh levels. Eventually, in the upper saltmarsh, the smaller *S. patens* becomes the dominant species (Bertness 1991; Figure 58).

*Spartina maritima* grows at a lower elevation than most other *Spartina* spp and in north-west Spain colonises from rhizome fragments. Thereafter, clones grow horizontally creating circular patches on bare mud (Sánchez et al. 2001). In the estuaries of Portugal and Spain it is quite robust (Figure 59).

In southern England, where the species reaches its northern limit, it was plentiful in the estuaries around the south and south east in the late 1800s, early 1900s (Marchant & Goodman 1969). It still survives in a few localities in south-east

**Figure 58** *Spartina alterniflora* saltmarsh on the east coast of America, Chincoteague National Wildlife Refuge. Saltmeadow Cord-grass (*S. patens*) lies flat in the foreground of the photograph

## 9.1 Introduction

**Figure 59** *Spartina maritima* clones in the Faro Estuary, Portugal

England although it is less robust than the same species, further south. The expansion of the more robust hybrid, *Spartina anglica* and possible climate change are amongst the reasons cited for its demise (Rodwell ed. 2000). Other species, including the hybrids, show similar patterns of colonisation.

### 9.1.2 Hybridisation

In North America, particularly on the east coast of the USA, *Spartina* is a dominant component of the extensive saltmarshes occurring there. Two species are important, *Spartina alterniflora* (Figure 58) and the smaller *S. patens*. The first species, in particular, readily hybridises with other species, and this has had a significant impact on saltmarshes throughout the world.

The first recorded occurrence of hybridisation took place on the south coast of England in about 1816. The native *Spartina maritima* crossed with *S. alterniflora*, introduced by shipping from the USA, accidentally into Southampton Water (England). This hybridisation produced a sterile plant *S. X townsendii* (Goodman 1969; Goodman et al. 1969). Following a doubling of its chromosomes, a fertile species *S. anglica* appeared and expanded rapidly (Marchant 1967). This event has been widely studied and reported. Several publications summarise the biology, origins, spread and impact of the species (e.g. Doody 1984; Gray & Benham 1990; Adam 1990, pp. 78–87).

In south-west France there may have been a second hybridisation event, involving the same parent species, but with different parental (nuclear) genotypes (Baumel et al. 2003). This plant does not appear to have had the same history as *Spartina anglica* and remains as a small isolated clone.

Hybridisation also occurs in other parts of the world, in San Francisco Bay, the native *Spartina foliosa* hybridised with the introduced *S. alterniflora*, colonising the Bay's tidal mudflats and marshes to the detriment of the former species (Callaway & Josselyn 1992; Daehler & Strong 1997a). Also, in San Francisco Bay, Dense-flower Cord-grass (*S. densiflora*) has recently hybridised with *S. alterniflora* (Ayres & Lee 2004).

The tidal range for *S. alterniflora* is also wide and varies throughout the world. In the USA, this variation is attributed to differences in mean tidal range (MTR). At the same time, the *S. alterniflora* zone expands with increasing tidal amplitude (McKee & Patrick 1988). It has the potential to grow from the Mean Higher High Water (MHHW) to approximately 1m below Mean Low Lower Water (MLLW) as seen in Willapa Bay, Washington (Sayce 1988). In San Francisco Bay, hybrids between the introduced *Spartina alterniflora* and the native California Cord-grass (*S. foliosa*) occur and grow at both lower and higher elevations than the native species. It is also more prolific than the hybrid (*S. anglica*) and the other non-native introduced species *S. densiflora* and *S. patens* present in the Bay (Ayres et al. 2004).

Hybridisation appears to give a competitive advantage over the original parents. In the UK, where the earliest recorded hybridisation between *Spartina alterniflora* and *S. maritima* took place, the resulting *Spartina anglica* has the ability to colonise almost any level in the tidal range. This included the 'absolute seaward limit of saltmarsh growth' to the landward limit of 'high water equinoctial tides'. It tolerates 'saline to brackish water conditions' (Ranwell 1972). At the lower tidal levels this new species was able to occupy a niche with limited competition and then only by scattered plants of annual *Salicornia* spp. (Gray et al. 1990).

In the USA, the *S. foliosa* X *S. alterniflora* hybrid in San Francisco Bay appears to be more robust than the native species, with superior seed set and siring abilities. This results in proliferation of hybrid clones capable of rapid expansion (Ayres et al. 2004). *S. densiflora*, a native of South America, is also tolerant of a wide range of conditions. It has not only successfully invaded the west coast of the USA, but also Spain and Morocco (Bortolus 2006). It is uncertain if a hybrid of this species, such as the one found in San Francisco, will create highly aggressive colonisers of tidal flats, similar to those of introduced *S. alterniflora*.

## 9.1.3 Pattern of Invasion

There are two key invasive species now: the native *Spartina alterniflora* and the hybrid *S. anglica*, included on the Global Invasive Species Database

## 9.1 Introduction

(see http://issg.org/database/welcome/). Their introduction, both accidentally and deliberately throughout the world, has lead to extensive colonisation of tidal mudflats. The nature of colonisation follows a similar pattern and is initially from seed or vegetative shoots. Towards their northern limit the number and viability of seeds depends on climatic factors, with the best seed set taking place in warmer years.

In some locations, they only spread slowly by vegetative growth following the first phase of establishment. This may last for many years with little or no lateral expansion. For the first 50 years the population of *Spartina alterniflora*, accidentally introduced to Willapa Bay in 1894 and only identified in the 1940s when it flowered, changed little. However, between 1945 and 1988, the species established itself throughout the bay (Sayce 1988). This rapid expansion appears to have occurred in favourable (warmer) years, with the production of fertile seed.

Other non-native *Spartina* species establish themselves in similar ways. *Spartina densiflora* in California invaded the upper levels of tidal flats mainly by vegetative spread, following seedling establishment (Kittelson & Boyd 1997).

Rapid invasion does not happen in all cases. In California, *Spartina anglica* had not spread beyond its original 1970s introduction site in 30 years. *Spartina densiflora* has spread to cover only 5 ha at 3 sites in the Central Bay over the same time period. *Spartina patens* had similarly only expanded from 2 plants in 1970 to 42 plants at one site in Suisun Bay (Ayres et al. 2004).

In other situations, expansion and contraction followed by reestablishment can take place. *Spartina townsendii* colonised the tidal flats on the eastern shore of Skallingen, Denmark resulting in a 'large area of coherent vegetation' from 1954 to 1964. 'Die back' occurred in the area of coherent vegetation from 1976 and 1988 only to be recolonised by 1995. To seaward, a series of circular patches stretched into the lower tidal area, the limits of which remained more or less stable throughout (Vinther et al. 2001).

The eroding saltmarshes of south-east England also have local examples where accretion takes place in sheltered locations such as the Blackwater Estuary. An example of this (Figure 60) also shows the typical process of colonisation, whereby clumps of *Spartina anglica* arise by vegetative growth, following seedling establishment.

### 9.1.4 Rates of Sedimentation

Worldwide vertical sedimentation rates of 20–80 mm per annum occur. Rates of between 100 and 120 mm per annum for *Spartina* were recorded for Bridgwater Bay, Somerset, in south-west England (Ranwell 1964). In exceptional circumstances, over short periods and in rapidly accreting saltmarsh, these can be as high as 200 mm per annum (Ranwell 1967). These compare with a range of 2 and 10 mm generally for middle and upper saltmarshes in Europe and eastern America (Ranwell 1964), in line with the 2.5 and 4.7 mm per year on a long established *Spartina alterniflora* saltmarsh, on the east coast of America (Flessa et al. 1977). Very similar accretion rates of between 2.4 and 4.8 mm per annum occurred in the American Pacific Northwest (Thom 1992).

**Figure 60** Coalescing clumps of *Spartina anglica* in the Blackwater Estuary, Essex, England

Sediment rates vary considerably depending on tidal range, sediment availability, and rate of compaction. Time of year is also important with rates being generally highest in the autumn (Ranwell 1964). Even at the lower end of the scale, *Spartina anglica* saltmarsh accretes at a rate that is some 10 times that of a typical 'natural' saltmarsh.

## 9.2 World Domination

Once the sterile *Spartina townsendii* had doubled its chromosomes and become fertile, the new species *S. anglica* showed a rapid expansion along the south coast of England. The plant first attracted attention when Lord Montagu of Beaulieu referred to the advance of *Spartina townsendii* (rice grass) over the tidal flat until it covered several thousand acres, in his evidence to a Royal Commission on Coastal Erosion in 1911 (Carey & Oliver 1918). Its ability to stabilise mud flats, led to its promotion for erosion control and as a land reclaiming agent (Oliver 1925). Plantings took place extensively in the UK (Goodman et al. 1959), with Poole Harbour on the south coast of England being the main source of material. Estimates of the overall area of *Spartina anglica* by 1967 amounted to 12,205 ha in 86 sites (Hubbard & Stebbings 1967).

Active promotion of the properties of the plant resulted in the export of considerable quantities of both seeds and plants from Poole Harbour (Ranwell

## 9.2 World Domination

1967) to many parts of the world. In addition to material from Poole Harbour, a Mr J Bryce sent other material from Essex. For example, in 1927/28, he exported seed to New Zealand and much later (between 1947 and 1955) plants (Partridge 1987). The result of this enthusiastic promotion resulted in a considerable increase in the world resources of this plant.

### 9.2.1 World Resources

Although not all the material successfully established (some seed was sterile) and conditions were not always suitable, by the early 1960s, the area of *Spartina townsendii* (s.l.) recorded worldwide was between 21,000 and 27,700 ha (Ranwell 1967). In addition to the large area in Great Britain, this included substantial areas in other European countries as well as smaller areas in Australia, New Zealand and the USA (Table 20).

**Table 20** Dates refer to the first appearance (or known introduction). Area estimates are of *Spartina* with ground cover of >50% and are very approximate (derived from Ranwell 1967)

| Country | Date of origin | Area (ha) |
| --- | --- | --- |
| Ireland | 1925 | 200–400 |
| Denmark | 1931 | 500 |
| Germany | 1927 | 400–800 |
| Netherlands | 1924 | 4000–5800 |
| France | 1906 | 4000–8000 |
| Australia | 1930 | 10–20 |
| Tasmania | 1927 | 20–40 |
| New Zealand | 1913 | 20–40 |
| USA | 1960 | <1 |

There were some early misgivings about the likely long-term effects of this rapid colonisation, 'whether the result (of *Spartina* establishment) will in the end be beneficial, or to the contrary will depend greatly on local conditions' (Stapf 1908). However, the general perception in these early days was that it is a great asset for sea defence, land reclamation and to a lesser extent animal fodder.

### 9.2.2 Spread in China, Australia and New Zealand

There are no native species of *Spartina* in China, Australia or New Zealand. Two species of *Spartina* (*S. anglica* and *S. alterniflora*) spread rapidly following their introduction to China in 1963 and 1979 respectively (Chung 1990; Bixing & Philips 2006). By 1985, the area of *S. anglica* reached 36,000 ha in 18 counties (Chung 1990). In North Jiangsu, *Spartina alterniflora*, covered some 410 km out of a coastal length of 954 km, with a maximum width over 4 km (Zhang et al. 2004). In China as a whole, the area increased from 260 ha in 1985 to 112,000 ha in 2002 (An et al. 2004). A review of the impact of the species over a 30-year period suggested that this colonisation was positive for partial control of siltation, fodder, nesting and feeding grounds for migratory birds, increased estuary productivity and in papermaking. *S. alterniflora* extracts provide additives for soda water, beer, milk, wine, tea and bathing lotions and have been trialled for their medicinal effects (Chung 1990; Chung 1993; Qin et al. 1997).

In Australia *Spartina anglica* infestations are found in the southern States of Tasmania and Victoria. In Tasmania, its introduction early in the nineteenth century was for the potential benefits in sea defence. It had invaded seven regions of Tasmania's coastal zone, occupying nearly 600 ha by 1997. Two sites, the River Tamar with 415 ha and the Rubicon estuary with 135 ha, represent a very small percentage of its potential habitat (Hedge 2002). In Victoria the estimated area of *S. anglica* was 186 ha (Hedge et al. 1997).

In New Zealand, expansion began in 1913 with the introduction of *Spartina townsendii*, which by 1952, in at least one site, the New River Estuary, Invercargill, had expanded to 40 ha (Partridge 1987). The maximum rate of spread of vegetation was 5.3 m per annum (Lee and Partridge 1983). Following the further introduction of *S. anglica*, by 1973, 90 ha of saltmarsh meadow and 250 ha, of scattered clumps were present, covering 15% of the mudflat (Hubbard & Partridge 1981).

### 9.2.3 USA, Washington State and San Francisco Bay

In the USA, there are three principal native species, two distributed on the east coast (*Spartina alterniflora* and *S. patens*) and on the south Pacific coast (*S. foliosa*). There are four introduced species:

1. *Spartina alterniflora* dominant, mainly Willapa Bay;
2. *Spartina anglica* widespread, dominant in Puget Sound;
3. *Spartina patens* small populations;
4. *Spartina densiflora* small populations first discovered in 2001 (Figure 61).

Initially non-native *Spartina townsendii* covered less than 1 ha (Ranwell 1967). Today the non-native invaders occur on the west coast where there are four principal sites (Figure 61). In Puget Sound, Washington State, *Spartina anglica* was introduced deliberately for shore stabilisation and as potential feed for cattle in the 1960s. Since then, in 36 years the plant has successfully invaded 73 sites, affecting 3,311 ha of marine intertidal habitat (Hacker et al. 2001). Also in Washington State *Spartina alterniflora*, accidentally introduced to Willapa Bay in the 1880s, had by the 1950s, expanded to cover about 10% of tidal flats. It spread

## 9.3 Changing Perceptions

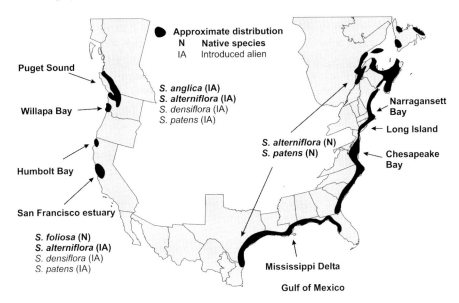

**Figure 61** Approximate distribution of the main native and introduced *Spartina spp.* in North America. **N** – native to area; **IA** – introduced alien

laterally at a rate of 0.79 m per annum (Feist & Simenstad 2000). Given the uniform slope of the intertidal flats of Willapa Bay, the potential area for colonisation is 66% or approximately 12,600 ha of the intertidal area (Sayce 1988). Between 1995 and 2000 the total area of non-native *Spartina anglica* and *S. alterniflora* in Washington State was 8,093 ha (Hedge et al. 2003).

For more information, see the Willapa Bay web site dedicated to providing information on *Spartina* spp. spread and control (see http://friendsofwillaparefuge.org/spartina.htm). The Washington State Department of Agriculture, Noxious Weed Control Board web site provides more general information on *Spartina* spp. (see http://www.nwcb.wa.gov/). It includes useful summaries of the species of *Spartina* found in the state.

In San Francisco Bay, hybrids of *Spartina alterniflora* and *S. foliosa* are the most numerous exotic species and spread rapidly (Strong & Ayres 2005). Three other non-native species are also present *S. anglica*, *S. patens* and *S. densiflora*, which together occupy approximately 5 ha. The total coverage of hybrids and other exotic species was 195 ha, slightly less than 1% of the area of tidal flats, although the potential area for colonisation was much greater (Ayres et al. 2004).

## 9.3 Changing Perceptions

The success of introduced *Spartina anglica* and *S. alterniflora*, hailed as a boon for sea defence, land claim and other human uses soon became questionable. The first suggestion that this rapid expansion might not always be beneficial came when the spread of *Spartina townsendii* (Oliver 1925) coincided with a loss of populations of *Zoster marina* in England in the 1920s and 1930s. Although there was no direct evidence of 'cause and effect', there are a number of accounts, referred to by Adam (1990) of *Spartina anglica* replacing native *Zostera* spp. It seems likely that *Spartina anglica* was the lucky recipient of the available niche, created by the loss of *Zostera marina* through a 'wasting' decease (Davison & Hughes 1998), rather than the agent of its demise. Experience from the UK suggested that, from a nature conservation perspective at least, the overall conclusion is that '*Spartina* provides a threat in estuaries of high wildlife interest, both to bird populations and to natural saltmarsh succession' (Doody 1984). However, it is the implications for bird populations feeding on exposed tidal flats, which was the principal cause of concern in the UK.

## 9.3.1 Impacts on Bird Populations in the UK

Concern about the rapid expansion of *Spartina anglica* and its effect on wintering waterfowl was one of the driving forces for a *Spartina* meeting held in Liverpool in 1982 (Doody 1984). This publication included papers on the origins, history and spread of the hybrid species *Spartina anglica*. The meeting discussed the implications of the spread for waders in the Dyfi Estuary in Wales (Davis & Moss 1984) and on invertebrates and shorebird populations at Lindisfarne National Nature Reserve, Northumberland, north-east England (Millard & Evans 1984). It was clear from both studies that the spread of *Spartina* reduces the area of upper open intertidal flats and with it the most profitable feeding zone for wading birds.

Lower down the shoreline *Spartina* may also have restricted the areas of *Zostera*, an important food for the Brent Goose and Wigeon. A more detailed study of the situation at Lindisfarne showed that, of all the factors potentially affecting wintering Brent Geese and Wigeon, the loss of the upper shore to *Spartina anglica* was more significant than sea-level rise and intermediate loss of *Zostera* spp. (Percival et al. 1998). These concerns resulted in several attempts to control the species at this site (Corkhill 1984; Frid et al. 1999). Further examination of these concerns, as they affected the wading bird, Dunlin in British estuaries, showed a negative correlation with *Spartina* expansion. However, the evidence that there was a causal link was not conclusive (Goss-Custard & Moser 1988).

## 9.3.2 Impacts on Amenity Beaches, North-West England

Changes associated with *Spartina* also have the potential to impact on recreational use. The Cheshire shore of the Dee Estuary, UK, provides an illustration of the extent of change and implications for recreational interests. Here, the marsh front advanced along the shore of the estuary, partly because of the growth and establishment of *Spartina anglica* (Taylor & Burrows 1968). The conversion of the sandy beach to a saltmarsh resulted in the loss of amenity beaches along the shore. This resulted in a request from the local authority to investigate the biology and control of the species (Taylor & Burrows 1968). By the mid-1990s the saltmarsh was 1.2 km wide in front of Parkgate; a sandy beach covered by most tides in 1939 (Figure 40; Pye 1996).

A similar problem arose on the south shore of the Ribble Estuary, further north along the coast of Merseyside. Here, Sefton Municipal Borough Council sought to control the expansion of *Spartina anglica* onto amenity beaches using the herbicide 'dalapon' (Robinson 1984; Truscott 1984).

### 9.3.3 Problems in the USA

At the height of the invasion of estuaries in Washington State, USA, dense swards of single species replaced natural vegetation. In the process, they destroyed important migratory shorebird and waterfowl habitat, increased the threat of flooding and severely affected the state's shellfish industry (Murphy 2005). These and other possible impacts (Table 21) led to increasing concerns about the invasion of *Spartina alterniflora* in the estuaries of the State.

**Table 21** Potential affects of *Spartina alterniflora* spread in Washington State (adapted from Callaway & Josselyn 1992)

| Possible impact | Cause |
| --- | --- |
| Competitive replacement of native plants | Higher seed production & germination; higher vegetative production |
| Effects of sedimentation | Greater stem densities, larger & more rigid stems |
| Changes in available detritus | Differences in quantity & quality of detritus |
| Decreased bottom-dwelling algae production | Lower light levels beneath *Spartina* canopy |
| Increased wrack deposition & disturbance to upper marsh | Greater stem production & subsequent deposition in high marsh |
| Changes in habitats for native wetland animals | Greater stem densities |
| Changes in bottom-dwelling invertebrate populations | Higher root densities & lower intertidal distribution |
| Loss of shorebird & wading bird foraging areas | Lower intertidal distribution |

Similar concerns exist in San Francisco Bay, where the plant which was only introduced in 1970, has spread rapidly right around the bay. Whilst the area of invasion represents only 1% of the tidal flats, the potential for further colonisation is problematic, not least for the survival of the native *S. foliosa* (Ayres et al. 2004). Some of the impacts of invasive *Spartina* highlighted for San Francisco Bay, USA, are:

- Loss of biodiversity as a result of the competition with native flora, including *S. foliosa* and *Salicornia virginica;*
- Hybridisation with native *S. foliosa;*
- Loss of mudflat and channel habitat;
- Loss of foraging and nesting habitat for numerous shorebirds, where the potential for loss is high (Stralberg et al. 2004) and waterfowl, including the endangered California clapper rail;

- Change in macro invertebrates (Neira et al. 2005);
- Clogging flood channels;
- Increasing rate of sedimentation and marsh elevation.

The San Francisco Estuary Invasive *Spartina* Project web site, (see http://spartina.org/index.htm) provides information that is more detailed.

### 9.3.4 Studies Elsewhere

In Victoria, Australia, concern about the spread of *Spartina anglica* began in 1991 when the Department of Natural Resources and Environment initiated a study to:

- Map the spread of the species;
- Raise the profile of the issue through workshops;
- Assess methods of control (Williamson 1995).

In February 1995, a *Spartina* Control Project for the area, containing approximately 98% of Victoria's estimated 186 ha of *Spartina* infestations, was initiated (Hedge & Kriwoken 1997).

New Zealand experienced much lower sediment accretion rates for *Spartina anglica* than elsewhere. 12 mm per annum in dense swards on a muddy substrate, to as little as 3 mm per annum on sandy substrates, with turbulent water in one estuary. The potential for loss of biological value still led to concern and attempts at control (Lee & Partridge 1983). The ecological, social and economic costs associated with its continued spread in Tasmania have resulted in the development of a management programme supporting eradication and control (Kriwoken & Hedge 2000).

In China, the apparent benefits attributed to *Spartina anglica*, resulted in the introduction, in 1979, of *S. alterniflora*. However, this species, brought in to check erosion, had spread to such an extent that it is 'choking estuaries, and crowding out native species such as the bulrush, *Scirpus mariqueter* together with its rich diversity of bird species (Chen et al. 2004) and reducing feed and habitat for fish and migratory birds'. The recent increase in trade with the USA has led to greater concern for the impact of alien species generally in China (Yan et al. 2000). Amongst these species is *S. alterniflora* listed by the Chinese Government on a 'Black List' of species, which should not be imported into the country (Normile 2004).

## 9.4 Methods of Control

In the early days, any thought that the control of *Spartina* spp. would become a major issue would have been an anathema to many people. This is especially true for those concerned with coastal erosion protection or land claim. However, it is clear from the examples described above that the hybrid *Spartina anglica* and American species, notably *S. alterniflora*, when outside their native range, can be aggressive invaders.

9.4 Methods of Control

This has resulted in a change in the perception of the value of the plant from being a 'friend' to a 'foe', where habitat loss and degradation are key issues. As a result, the 'route to the restoration' involves reversing the encroachment of the species onto open intertidal sand and mud flats. Some of the methods are considered next.

## 9.4.1 Herbicides

A common method of control is spraying with herbicides. Marketed under a number of different commercial names, they can represent the most effective means of control. However, depending on their toxicity to humans, effects on the environment generally, and on non-target species, they are closely regulated in most countries. Making specific recommendations is thus not possible. What follows is a brief review of some of the methods and their efficacy from published material around the world.

In Great Britain, attempts to control the species included the use of a variety of herbicides at Lindisfarne National Nature Reserve, Northumberland, northern England (Corkhill 1984). This included Dalapon (sodium dichloropropionate), also used to control *Spartina anglica* on amenity beaches (Truscott 1984). The treatments met with varying degrees of success. It was particularly effective on the amenity beach, where it achieved 99% kill after three applications. Although 90–100% kill occurred in some trial areas after 5 years most of the *Spartina* continued to grow and expand (Corkhill 1984). There was evidence of a return of some of the bird populations using the site. However, there was no clear indication of the long-term efficacy of the treatment (Evans 1984).

The herbicide Roundup PRO, based on glyphosate, was much less effective at Lindisfarne National Nature Reserve, achieving only a 50% kill in initial applications (Corkhill 1984). Reviews that are more recent suggest that in Washington State, USA, where Dalapon is no longer used, the only alternative, the Rodeo formulation of glyphosate, showed very variable effectiveness (Hedge et al. 2003). The annex (p. 123), gives a summary of the properties of Rodeo (Department of the Interior, US Fish and Wildlife Service, Willapa National Wildlife Refuge 1997). Alternatives that may prove more effective include Imazapyr (Patten 2002).

Elsewhere, such as Australia, other herbicides also proved successful. Other than for small infestations, the only herbicide which proved to be effective was Fusilade (active constituent: 212 g/L fluazifop-P present as butyl ester (FPB)) providing up to 100% kill, with 'acceptable' environmental side effects (Hedge & Kriwoken 1997). In Tasmania, the same chemical was the only suitable and effective herbicide to control 'rice grass' (Hedge 2002). In New Zealand, chemical treatment was by far the most effective. Here, treatment is mostly with the herbicide Gallant (Haloxyfop), either Dalapon/Weedazol or Roundup (Shaw & Gosling 1997).

In China, the recent change in approach to *Spartina* spp. invasion has resulted in work to find a weed-killer suitable for treatment. Studies identified an herbicide called micaojing, which killed all of the above ground parts of *Spartina* app. within

30 days. It is reported that the below ground parts are also susceptible, being killed within 60 days. Micaojing appears to be harmless to wild animals including clams, tuna and prawns and disappears from the environment in 30 days (Jian et al. 2005). To date there is no common agreement on the best herbicide. Much depends on the physical conditions at an individual site.

### 9.4.2  Physical/Mechanical Control

A detailed Environmental Assessment (EA) for the control of *Spartina alterniflora* on Willapa National Wildlife Refuge, Washington State from 1996, considered a variety of physical/mechanical methods. These included commercial harvesting, trampling/crushing, excavation, scraping, ploughing/rotovating, dewatering/draining, flooding/inundating, burning, steaming, covering, use of laser beams and freezing roots. Of those investigated, hand pulling or 'pushing' plants into the mud worked, but only on young seedlings. Even then, it is important to remove both above and below ground parts of the plant. Cutting, mowing alone, or burning needed several treatments and only eliminated infestations at high cost (Department of the Interior, US Fish and Wildlife Service, Willapa National Wildlife Refuge 1997).

At Lindisfarne National Nature Reserve, chemical control all but ceased. Instead, in the 1990s mechanical trials were undertaken using a machine, which 'buried' the plants. Burying using a rotoburying machine at Lindisfarne killed over 95% *Spartina anglica* after two years (Denny & Anderson 1999). Use of a light-weight tracked vehicle driven repeatedly over *S. anglica* resulted in a stem density approximately half that for untreated plots (Frid et al. 1999).

In Victoria, Australia, methods included slashing, burning, sluicing, digging, smothering and herbicides. Except for small newly formed infestations only herbicides proved to be effective (Hedge & Kriwoken 1997). Smothering techniques also proved less than successful, being very labour-intensive, suitable for small areas only and susceptible to damage by storms and from vandalism (Hedge 2002).

### 9.4.3  Grazing

Grazing clearly has an effect on *Spartina* swards, as it does on saltmarsh more generally (Section 7.3). Grazing, more than mowing or cutting, is likely to reduce seed set and hence expansion. However, it is unlikely to eliminate the saltmarsh from the tidal flats and does not appear to have been used as a control mechanism.

### 9.4.4  Biological Control

Greenhouse experiments found that *Spartina alterniflora* clones became stressed or killed by moderate populations of *Prokelisia marginata*, a Homopteran leafhopper common to the home range of *S. alterniflora* (Daehler & Strong, 1997b). A greenhouse population of *S. anglica* introduced to Puget Sound in Washington was also vulnerable to high populations of planthoppers from California (Wu et al. 1999). Thus, early results suggest the most effective biological control so far appears to be from *Prokelisia marginata*. This feeds on *Spartina* fluids, by piercing the leaf (Hedge et al. 2003).

## 9.4.5 Summary of Control Measures

Throughout the world, there have been many attempts to control *Spartina*. These are mostly costly and/or ineffective (Table 22).

Table 22 A summaries of control measures, taken from various sources. See, for example, Hammond & Cooper (2002); the San Francisco Invasive *Spartina* Project, which provides a wealth of information as well as links to many other sites (see http://www.spartina.org/index.htm); Hedge (2003) and Roberts & Pullin (2006)

| Method | Effectiveness | Advantages & disadvantages |
|---|---|---|
| Herbicides (Dalapon, Glyphosate) | Can be effective although Dalapon is difficult to obtain and Glyphosate and other herbicides not fully trialled | Requires continual treatment. Relatively expensive |
| Digging | Partially and on a small scale (mainly seedlings) | Labour-intensive and costly on a large scale |
| Dyking and inundation | Partially effective in preventing spread | Costly and damaging to other saltmarsh communities |
| Bulldozing (removal of surface) | Ineffective | Potential damage to mud surface |
| Rotovating & harrowing | Counterproductive | Greater propagation from broken rhizomes |
| Burying (ploughing & rotoburying) | Effective if plants are covered. Effective for up to four years | Difficulties of access |
| Crushing | Partially effective | Requires repeat treatment, vehicular access difficulties |
| Burning | Ineffective | Impractical |
| Grazing | Prevents seedling production and hence can restrict spread. Cost-effective | Increases shoot density, no reduction in clumps |
| Mowing | Prevents seedling production, and hence can restrict spread. Can be labour-intensive | Can increase shoot density, no reduction in clumps. Requires continuing treatment |
| Covering (black plastic) | Partly effective on a small scale | Difficult to keep plastic in place |
| Biological control | Can be effective on an individual site basis. Avoids use of chemicals | Involves introduction of alien species. Experimental |

## 9.5 *Spartina* spp. Friend or Foe?

The several species of *Spartina* pose an interesting dilemma for the conservationist and coastal manager alike. Native species of *Spartina* in their own environment provide a significant contribution to the functioning of the coastal ecosystem in which they occur. They also form an important component of many restoration schemes, especially in the USA (Section 6.2.5). However, aggressive hybrids and species outside their normal range can cause significant environment problems, as has been outlined in this chapter.

It is generally accepted that native *Spartina* spp., which form part of a natural succession, will have values associated with accreting State 3 or State 2 saltmarsh. It is also true that, promoting saltmarsh accretion through the introduction of *Spartina anglica* and other *Spartina* spp. is very successful. This has created a backlash in many parts of the world with the resulting 'demonisation' of the plant and attempts to eradicate it (Section 9.4).

However, factors such as the apparently natural 'die back' and a reappraisal of the role of *Spartina anglica* in the 'natural' succession have raised questions over eradication as a form of restoration, especially in the UK (Lacambra et al. 2004). The rest of this chapter looks at the pros and cons of those species of *Spartina* currently perceived as being a threat to one or more environmental values.

### 9.5.1 Control – Concerns and Costs

Overall, *Spartina* control has proved to be a costly and complex process (Hedge et al. 2003). It also appears that many of the methods employed are inefficient. Herbicide treatment remains the most effective control mechanism. However, even here 'approved' chemicals are not always completely successful and may require several treatments for near 100% eradication. Reinvasion is always possible and it seems likely in most areas that total eradication is not possible, except where small-scale local invasions are involved. Many of the apparently most effective herbicides (used in Tasmania, New Zealand and China, Section 9.4.1), are largely untested, elsewhere. Regulatory approval, in the face of concerns about toxicity to other non-target species and possibility of persistence in the environment, will continue to make widespread acceptance of their use difficult.

In Washington State, some of the most intensive and expensive control programmes have been undertaken. Despite five years of treatment between 1995 and 2000, which covered an average of 15% of the *Spartina* infestation, the total area increased considerably (Hedge et al. 2003). Following an increasing effort for three years up to 2005, there were still 2,550 solid hectares in Willapa Bay in 2004 and 223 solid hectares in Puget Sound. Nearly 80% of the former area and 95% of the latter were treated in 2005. The cost of control is impressive, with a budget of more than $1.5 million allocated for 2006 (Murphy 2005). Details of the programme are contained in annual reports from 1998 to 2006, (see http://agr.wa.gov/PlantsInsects/Weeds/Spartina/default.htm).

## 9.5.2 'Natural Die Back'

It is perhaps not surprising, given the often rapid and extensive colonisation of *Spartina* described above, that control measures take place. 'Die back' is a phenomenon often used to describe *Spartina* spp. plants exhibiting reduced growth, which can result in death of individual plants and ultimately the loss of large swaths of *Spartina*. Soon after the rapid expansion of *Spartina townsendii* s.l., in the early 1900s, slowed in the 1950s, 'die back' appeared on the south coast of England.

Studies of the process identified two distinct forms 'edge die back' and 'die back' in and around 'pans' in the centre of the saltmarsh (Tubbs 1984). Losses at the edge of the saltmarsh were attributed to wave attack, whilst those in the centre seemed to be associated with water logging and soft-rotting of the apex of the rhizome (Goodman 1960). The process of expansion and retreat, described for the south coast of England, represents a typical pattern of change (e.g. Goodman et al. 1959; Gray & Pearson 1984). Langstone Harbour, in the Solent Estuary provides an illustration of the scale of the change. Erosion and slumping followed the expansion of *Spartina* in the first half of the twentieth Century, such that its area was considerably reduced by 1980 (Figure 62).

A similar pattern of change occurred subsequently in South Wales, and along much of the east coast including North Norfolk and in northern France and south-west Netherlands (Gray et al. 1997; Gray & Raybould 1997). 'Die back' also appears to occur naturally in inland areas of native *Spartina alterniflora* saltmarshes in Louisiana, representing a significant area of loss (Mendelssohn & McKee 1988). The Mississippi River Delta suffered a major and rapid loss in 2000.

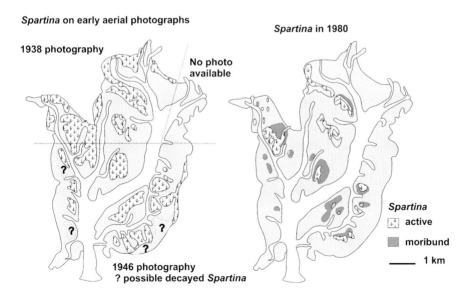

**Figure 62** Change in the area of *Spartina* in Langstone Harbour, redrawn from Haynes (1984)

However, these areas recovered relatively quickly (McKee et al. 2004). Large-scale 'die back', affecting approximately 800 ha of saltmarsh, occurred in Georgia, although again this was not long lasting (Ogburn & Alber 2006).

The invasion of *Spartina anglica* in China seems to have followed a similar pattern. After its introduction in 1963, it had spread to 36,000 ha by 1985, in 18 counties. However, by 2002 there were only a few small and scattered colonies in three counties, the majority having suffered extensive 'die back'. By contrast the spread of *S. alterniflora* continued. Following its introduction in 1979, it had spread to 260 ha by 1985, reaching 112,000 ha by 2002 (An et al. 2004).

The production of phytotoxins under anaerobic soil conditions created by poor drainage, or even to rising relative sea levels is one of the most frequently suggested causes. Water logging seems to be a key factor in increasing soil reduction and sulphide concentrations (McKee et al. 2004). The development of 'salt pans' in Australia is attributed to water logging of stands of *Spartina townsendii* (s.l.) (Boston 1983). Although there is no specific mention of 'die back' this could be a similar process. Whatever the mechanism, it is possible to view the expansion and subsequent retreat as a natural process, whereby a new species occupies a previously unoccupied niche and has paved the way for its own destruction (Gray et al. 1991).

The Langstone Harbour example also illustrates the nature of the change. As saltmarsh is lost, dense growths of algae cover the mudflats, resulting from an increase in eutrophication, partly brought about by decaying *Spartina* (Figure 63).

**Figure 63** Growth of algal mats in Langstone Harbour, following the loss of *Spartina*, photograph taken in 1980. *Spartina* remains present as isolated clumps in the middle distance

Once 'die back' occurs there is little evidence of reinvasion, suggesting that conditions remain unsuitable for some time. A nature conservation assessment of Langstone Harbour in January 2007 concluded that in approximately 50% of the intertidal area, the condition of the site in three of the main units was 'unfavourable', due to the continued erosion of saltmarsh. Information from the Natural England web site, Site of Special Scientific Interest Assessment Report, January 2007 (see http://www.english-nature.org.uk/Special/sssi/sssi_details.cfm?sssi_id=1001182).

Holes Bay, Poole Harbour, on the south coast of England, provides an indication of the timing of these changes. *Spartina* arrived in 1899 and expanded relatively rapidly to produce swards occupying 208 ha, more than 60% of the intertidal mudflats by 1924. By 1972 it had retreated to less than half the area, by 1994 to less than a third (63 ha) (Gray & Raybould 1997).

## 9.5.3 Changing Patterns of Invasion – Great Britain

The pattern of change described for individual sites appears to extend to a wider geographical area though with distinct regional differences. The estimated 12,205 ha of *Spartina anglica* in 86 sites in England and Wales in the 1960s (Hubbard & Stebbings 1967) had apparently fallen to 6,950 ha by the end of the decade (Way 1990, quoted in Lacambra et al. 2004). By the 1990s, approximately 10,000 ha of *Spartina anglica* represented nearly 25% of the total saltmarsh in Great Britain (Gray et al. 1997). A review of the status of *Spartina* in Great Britain in the late 1980s helps to explain this apparent reversal of fortunes. Taking the area of *Spartina* in Hubbard and Stebbings (1967) as a starting point, it is possible to compare the changes taking place in different geographical areas (Table 23).

These figures are not directly comparable but support the view that an early rapid phase of expansion and retraction took place in the south. This reflects the stages in the growth, establishment and recession in different geographical areas. Despite an increase in the total number of sites on the south coast, there has been an overall reduction of 11% in the area of *Spartina* saltmarsh. This reduction is even more obvious on the east coast. The increase on the west coast is equally clear.

**Table 23** Areas of *Spartina anglica* as given for three geographical coastal areas in Great Britain. A literature search and limited survey provide the basis for the comparison (Charman 1990)

|  | South | East | West | Total |
| --- | --- | --- | --- | --- |
| Hubbard & Stebbings (1967) |  |  |  |  |
| Area of *Spartina* (ha) | 3,326 | 6,568 | 2,312 | 12,205 |
| Number of sites | 24 | 27 | 35 | 86 |
| Updated figures (Charman 1990) |  |  |  |  |
| Area of *Spartina* (ha) | 2,951 | 3,655 | 3,248 | 9,854 |
| Number of sites | 29 | 27 | 55 | 111 |
| New sites / old sites | (+6 −1) | (+1 −1) | (+20) | (+27 −2) |
| Change | −11% | −44% | +40% | −19% |

It seems that the rapid growth and subsequent 'die back' on the south coast has occurred at a later date on the east coast. Expansion continues on the north-west coast of England (see Section 9.5.4).

Analysis of data from two other sites in England helps provide further appreciation of these changes and the management response. The main expansion of *Spartina anglica* in Bridgwater Bay, Somerset, occurred between 1947 and 1971. There appears to have been a contraction from 1971 to 1982 and again by 1994, though at a slower rate. A survey in 1999 showed the boundary to be similar to that for 1994. This natural decline took place without any intervention to control the colonisation. At Lindisfarne, National Nature Reserve, Northumberland, the expansion of *S. anglica* was later still. Although present in the 1960s, its slow expansion did not cause alarm. However, after 1964 with the building up of the causeway road to Holy Island, a more rapid expansion took place. Due to the threat to wintering waterfowl, it was decided to control the plant by hand-pulling and digging in the 1970s, and later by the use of chemicals (Corkhill 1984). Despite these control measures, *Spartina* continues to spread (Lacambra et al. 2004).

### 9.5.4 Spartina in North-West England, a Case of Succession

A visit by the author to the island of South Walney in August 1981 proved to be of some interest in relation to the '*Spartina* story' in the UK. The extensive tidal flats east of the island were in the first phase of a remarkable transition. At the time, the area formed part of an extensive Morecambe Bay Site of Special Scientific Interest. It was also identified as a part of a Nature Conservation Review site (Ratcliffe 1977), which formed the basis for the selection of sites designated under the European Union, Habitats and Species Directive. The principal reasons for selection include sand and mudflats exposed at low tide, together with their feeding wintering waterfowl population (Table 24). The site is also important for its colonising *Salicornia europaea* and Atlantic saltmarshes.

The history of *Spartina anglica* expansion in the Morecambe Bay estuary is typical of the situation in the rest of Great Britain and many other parts of the world. It appears to have arrived naturally in the 1940s, from which time its population remained more or less stable. By the end of the 1960s, there were a few new sites but there had been only slow spread. By 1982, the species had established in a few localities, although most populations occurred as isolated clumps (Whiteside 1987). Outside the survey area, in the outer reaches of Morecambe Bay *Spartina* clumps were clearly visible off the eastern shore of South Walney Island (Figure 64) in August 1981 (Figure 65).

In 1985, these isolated patches had coalesced to form continuous swards (Figure 66). A survey of the whole shore in the lee of the island, at about the same time, estimated saltmarsh as covering 240ha (Burd 1989). Partly because of this and other changes in *Spartina* noted on the UK coastline, a symposium took place in Liverpool in November 1982, to review the status of *Spartina* (Doody 1984).

## 9.5 *Spartina* spp. Friend or Foe?

**Table 24** Principal wintering waterfowl using the tidal flats of Morecambe Bay Special Protection Area (designated under the European Union Birds Directive) as a percentage of the world, North West Europe (NW) and East Atlantic Flyway (EAF)

|  | Migratory >1% International Biogeographical Population |
|---|---|
| Pink-footed Goose | 4.1% world |
| Shelduck | 2.3% NW |
| Pintail | 3.8% NW |
| Oystercatcher | 6.0% EAF |
| Ringed Plover | 3.0% EAF |
| Grey Plover | 1.1% EAF |
| Knot | 8.3% EAF |
| Sanderling | 3.0% EAF |
| Dunlin | 4.3% EAF |
| Bar-tailed Godwit | 1.8% EAF |
| Curlew | 3.6% EAF |
| Redshank | 4.3% EAF |
| Turnstone | 2.5% EAF |

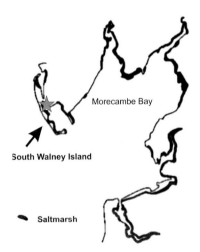

**Figure 64** The approximate location of the photographs taken from South Walney Island and shown in Figures 65–67 is indicated by a star

A further visit in 2005 showed how the upper levels of *Spartina* appeared to have begun to give way to a high-level Atlantic saltmarsh community, one of the features for which the site was designated under the European Habitats Directive (Figure 67). In this community, *Spartina* is much less dense and other species such as *Limonium vulgare*, *Triglochin maritima*, *Juncus maritimus* and *Atriplex portulacoides* appear in the sward. The presence of low-level cattle grazing (Figure 66) may have helped the successional process.

**Figure 65** August 1981 circular patches of *Spartina anglica* stretching onto the tidal mudflats of the Piel Channel Flats, Morecambe Bay, from South Walney Island

**Figure 66** August 1985 *Spartina anglica* meadow (compare with Figure 65)

**Figure 67** July 2005 similar view to Figures 65 and 66. There is abundant *Limonium vulgare* and other species characteristic of upper saltmarshes in the foreground

## 9.6 Conclusion

Promotion of the hybrid *Spartina anglica* as a land-reclaiming agent was one of several positive values attached to the species. The resulting introduction to estuaries around the world of this species, and *S. alterniflora*, has been remarkably successful. However, in recent years, a consensus has grown that suggests the positive economic benefits associated with the invasion of *Spartina* spp. no longer outweigh the largely negative environmental consequences, especially those associated with nature conservation. Even in China, despite its commercial use, recent reports indicate a change in attitude away from promoting its spread to one of control. Consequently, in many parts of the world, *Spartina* hybrids or native species outside their normal geographic range are subject to strategies designed to control or eradicate them.

### 9.6.1 Spartina anglica – A Natural Component of Saltmarshes in the UK and Ireland?

Given the concern around the world about the spread of *Spartina anglica*, it is perhaps surprising that the species is now considered an endemic 'native' in the

*Atlas of the British & Irish Flora* (Preston et al. 2002). This recent reclassification reflects a developing view that the species has been around long enough to become a 'natural' component of saltmarsh vegetation.

In an attempt to inform management policy on nature reserves and other protected areas, English Nature (Natural England) commissioned a review of the species. The detailed report includes a review of the situation generally, as well as providing a comparison between two different National Nature Reserves (NNRs) in England both with mudflats, important for wintering waterfowl. At Bridgwater Bay, NNR, despite rapid expansion of *Spartina anglica* on to the mudflats, there has been no control. By contrast, at Lindisfarne NNR, control has been taking place since 1970. At the former site, *Spartina* has stabilised, covering an area less than its maximum spread. At the latter site, it continues to expand (Lacambra et al. 2004). The limitation on expansion at Bridgwater Bay may result from the effects of wave action (Morley 1973).

These two different approaches show the importance of determining management on a site-by-site basis. The situation on the tidal flats of South Walney also reflects a differing view on the threat posed by rapid expansion of the species. In this case, a recent review of the nature conservation status of the saltmarsh and mudflat by Natural England considered 'condition' of the tidal saltmarshes at South Walney as 'favourable'. This was despite a recorded doubling of the area of *Spartina* saltmarsh in recent years. There have been no major attempts to control the species in this area despite the high value of the site for wintering waterfowl, many of which use the tidal sand and mud flats to feed on.

In Ireland, there is no indication that it is considered a problem, indeed where it occurs it is thought to enhance biodiversity (Curtis & Sheehy Skeffington 1998). By contrast, in Northern Ireland *Spartina anglica* is a major issue in several estuaries, notably Strangford Lough. Originally introduced in the 1940s, it spread to the point where it became a threat to wintering waterfowl populations. In the 1960s, an attempt to eradicate it using herbicides, failed (Hammond & Cooper 2002).

## 9.6.2 Friend or Foe

The question as to whether *Spartina* spp. is friend or foe, depends on a number of issues. In its natural geographical areas, it is valued for the protection it affords to both the hinterland and for species living in and around it. It provides many of the other values associated with saltmarsh in terms of productivity, acting as a pollution sink, etc. (Section 4.3.2) and nature conservation (Section 4.5). In the early days, whether as a hybrid or a native species outside its natural range, its values for coast protection and as an aid to land reclamation outweighed the environmental disbenefits. With the recognition of the environmental and nature conservation problems, such as the loss of intertidal feeding for wintering waterfowl, the view of the plant has changed. The '*Spartina* phenomenon' had become a problem requiring control or eradication (Doody 1990).

## 9.6 Conclusion

However, despite this it seems that control may not always be appropriate. The incidence of 'natural' change, notably 'die back', suggests that for *Spartina anglica*, leaving things alone may be all that is needed. The cost, the ineffective nature of some methods of control and the need for repeat treatments, also call it into question. Natural succession, in the UK at least, appears to be leading to the development of habitats of high nature conservation value in their own right at some sites. Although leaving things alone may result in the expansion of areas of saltmarsh at the expense of tidal flats, the overall nature conservation value appears to survive.

Thus, the loss of tidal flats to *Spartina* invasion will almost certainly illicit a first response similar to that adopted worldwide. Given the speed and scale of invasion and the known effects on many features of nature conservation and economic interests this is entirely understandable. However, this initial response requires modification in order to avoid costly and ineffective management.

In the Odiel Estuary, *S. densiflora* invasion may have helped create the large expanse of saltmarsh at this site. Today it dominates some 18% of the saltmarsh community (Mateos Naranjo et al. 2006) but elsewhere it is scattered throughout the vegetation (Figure 68). The saltmarshes form a significant component of this important site, which is a Natural Park, Ramsar Site and Special Protection Area designated under the European Union, Bird Directive.

There is no information on its original expansion at this site. The species probably first appeared following its accidental introduction from South America to the Gulf of Cádiz in the sixteenth century (Castillo et al. 2000). In the estuaries in north-west England, the recent invasion of *Spartina anglica* has not resulted in a clamour for its destruction, at least not on ornithological grounds. Although *S. densiflora* may represent a threat to saltmarshes and mudflats elsewhere in southern Europe, it is difficult to see what can or should be done to curtail its expansion. The Odiel Estuary is sufficiently large to support a rich bird fauna, including international important wintering waterfowl. Thus the appearance of this alien may have changed the nature of the estuary without destroying individual components including its birdlife.

The situation for some of the other alien species, especially *Spartina alterniflora* is less clear. The speed of invasion, its hybridisation with native species in the USA, the knock-on effects for nature conservation and impact on commercial interests may point to the need for control. However, for many of the sites, invasion is relatively recent. Given time, the situation in the UK suggests that invading *Spartina anglica*, at least, can become an acceptable component of the saltmarsh habitat. By allowing time, natural processes such as 'die back' or succession can lead to 'incorporation' of the species into the 'natural' habitat.

**Figure 68** *Spartina densiflora* in a matrix of *Sarcocornia perennis* with *Atriplex portulacoides*, Odiel Estuary, Spain

# Chapter 10
# Conclusions

## A Place for Saltmarsh and Everything in its Place

## 10.1 Introduction

This book has described saltmarshes, their development and position in the wider coastal ecosystems (Chapter 1). It has considered the extent of human influence from grazing management, through to enclosure for agriculture (arable and rice cultivation) and the development of infrastructure such as ports, harbours, industry, housing and roads (Chapter 2). Recognition that losses have been extensive, putting at risk both nature conservation and socio-economic interests, led to a change in the view of the role of saltmarshes (Chapter 3). Consideration of the values attached to different physical and biological (vegetated) 'states' (Chapter 4) led to the development of 'models' for habitat restoration and vegetation management. Included in this were discussions of the trends and trade-offs associated with moving between the different physical states (Chapter 5) and ways of achieving these (Chapter 6). Similar considerations apply in relation to grazing management dealt with in Chapters 7 and 8. An additional chapter considers the special case of *Spartina* spp. (Chapter 9).

This final chapter discusses the relationships between managed and restored saltmarshes and their role in the wider tidal environment. The way in which these activities redress the balance between ecosystem and human values of saltmarsh is considered. In particular, it looks at the role of the saltmarsh in maintaining the ecological balance, biological diversity and sustaining human use.

## 10.2 Time and Tide Wait for No One

To understand the relationship between 'natural' and human change we need to set a context. One of the more important factors affecting saltmarsh development is sea-level change. Sea levels have risen and fallen in response to changes in the coverage of the ice-sheets over millenniums. Since the last glacial maximum about 20,000 years ago, the sea level has risen, from its lowest point, by over 120 metres because of the melting of the ice. From some 15,000 years ago, there was a rapid

**Table 25** Estimated rates of global sea-level change during the Holocene. Detailed information on global warming and the implications for sea-level change; including the Intergovernmental Panel on Climate Change report referred to (Houghton et al. 2001)

| Period | Rate of change (mm/year) | Reference |
| --- | --- | --- |
| 15,000–6,000 years ago | 10 | Houghton et al. 2001 |
| 6,000 years ago to the present | 0.5 | Houghton et al. 2001 |
| 3,000 years ago to present | 0.1–0.2 | Houghton et al. 2001 |
| 1880–1980 | 1.8 | Douglas 1991* |

\* This estimate of 1.8 mm per year derives from 21 stations in nine oceanic regions avoiding tide gauge records, in areas of converging tectonic plates or with substantial uplift or subsidence due to 'post glacial rebound'

rise in sea level. This resulted in an inundation of the coast with the subsequent landward movement of coastal habitats including saltmarshes. From about 6,000 years ago to the present day rates of change slowed dramatically (Table 25). These estimates refer to average changes in global sea level, referred to as eustatic change. The terms 'eustasy' or 'eustatic' refer to changes in the volume of water in the oceans, usually resulting from changes in the global climate.

A reappraisal of these figures confirms the more recent general rate of 1.8 mm per year during a one hundred year period. Despite the fact that this appears to be an order of magnitude greater than in the previous 3,000 years, there is no discernable acceleration in the rate of sea-level rise over the last 100 years (Douglas 1997). Analysis of the relationship between foraminifera and plants in saltmarshes in the Gulf of Maine, however, indicate that since 1800, sea levels have risen by 0.3–0.4 m (c 1.5–2.0 mm per year) here at least. The results suggest this rise corresponds with regional climatic warming and thermal expansion of the Gulf of Maine and North Atlantic sea surface. Despite a temporary slowing during the mid nineteenth century, twentieth century rates are higher than at any time in the last 1,000 years and correspond with global warming trends (Gehrels et al. 2002).

Satellite data seems to confirm this. At a global scale altimetry data from the Topex/Poseidon satellite (1993–1998) shows a 3.2 ± 0.2 mm per year global mean sea-level rise, which is attributed to thermal expansion of the oceans (Cabanes et al. 2001). In the Mediterranean during 1993–1999, data from the same satellite indicates rates as high as 20 mm per year south-east of Crete (Cazenave et al. 2001) though smaller changes were observed elsewhere, including a decrease in the northern Ionian Sea.

The relative rate of sea-level movement on the coast depends on the relationship between changes in the level of the sea and the land levels. A number of factors depress or release the Earth's crust and result in changes in land levels, referred to as Isostatic change. The principal cause is the formation of ice-sheets, which depress the land surface. As the glaciers melt the land recovers with the removal of the weight of ice, a process called 'Isostatic post-glacial rebound'.

The true rate of sea-level change as it affects the coast, thus depends on changes in both the land and sea surfaces. This 'relative sea-level change' can be much greater than the sea-level change alone. Tide gauge records suggest that in Northern Europe between 1900 and 1985 sea levels have fallen. In areas of maximum crustal loading during the last Würm Glaciation, this has been by as much as 10 mm per year in the northern Baltic Sea. In unglaciated areas, such as southern England and parts of northern France, it is rising by as much as 6 mm per year (Aubrey & Emery 1993). In Scotland, a major ice-sheet covered the land and today the surface is rising by 1.6 mm per year. By contrast, Southern England remained unglaciated throughout the glacial period and sinks today at a rate of about 1.2 mm per year. (Shennan & Horton 2002). Other factors, which may influence land levels are sediment deposition, compaction and plate tectonics. The maximum rates are even greater in Japan where the rate of emergence of the land from the sea can be as high as 6 mm per year. In the same area, the rate of submergence in the south-east is up to 20 mm per year. These rates are due to tectonic trends caused by the convergence of two continental plates (Aubrey & Emery 1993).

### 10.2.1 Can Saltmarshes Keep Pace with Sea-Level Rise?

Against the background of long term global changes in relative sea level the physical conditions, tides and storms determine whether saltmarsh develops or not. Given an adequate supply of sediment, up to a point, saltmarsh can keep pace with sea-level rise. This requires that the rate of accretion minus any lowering of the surface due to compaction is greater than or equal to the rate of sea-level rise. In the early stages of colonisation, rates of vertical accretion can reach 10 mm per year or more (Section 4.2.1.), well in excess of the rate of sea-level rise in most areas. For mature saltmarshes there appears to be a limit above which sea-level rise outstrips saltmarsh growth. This limit is around 6 mm per year in the Wadden Sea (Bakker et al. 2005). Thus, even in areas where land levels are sinking and the maximum rates of relative sea-level rise approach 6 mm per year, as in parts of the southern North Sea, saltmarsh accretion can theoretically keep pace.

Why then are most of the world's coastlines in a state of erosion (Pilkey & Cooper 2004), and it appears that apart from areas where *Spartina* spp. is present (Chapter 9) this is also true for many saltmarshes? Part of the reason lies in the fact that erosion is a natural process in saltmarsh development (see Section 4.2.2). Thus, not all saltmarshes exhibiting erosion are undergoing irreversible change. Acceleration in sea-level rise may change the point at which a dynamic equilibrium occurs, causing a fall relative to the tidal frame. Whilst this may result in 'drowning', it causes a change in vegetation rather than a loss saltmarsh. It seems that areas where sediments are depleted are at most risk (Nicholls & Leatherman 1995; Neuhaus et al. 2001). In addition, many surviving saltmarshes lie in areas where massive enclosure has occurred. The fact that this prevents landward migration as sea levels rise (Section 3.5.2) further exacerbates the likelihood of erosion.

Predictions of sea-level rise derived from models prepared by the Hadley Centre in the UK, suggest that a rise of some 3.8 mm per annum is possible between 1990 and the 2080s. This could include significant wetland losses (saltmarsh, mangroves and intertidal areas) of up to 22%, which when added to the anthropogenic losses due to enclosure, could amount to as much as 70% in some places. These losses will influence many sectors and values, including food production (such as loss of nursery areas important for fisheries), flood and storm protection (storm surges will cause flooding further inland), waste treatment and nutrient cycling functions, and as habitat for wildlife. An accelerated sea-level rise could significantly worsen the already poor prognosis for coastal wetlands (Nicholls et al. 1999; Nicholls & Hoozemans 2005). Given this scenario, it seems saltmarsh restoration will become even more significant in the years ahead.

## 10.3 Saltmarshes and Saltmarsh Restoration

Chapter 4 provides a description of the values associated with saltmarshes, amongst which their coastal defence function is highly significant. The nature of the vegetation can influence the stability of the saltmarsh. However, other factors such as sediment availability, exposure to waves and extent of enclosure may be more significant. Preventing erosion is an accepted, practical way of helping to sustain most of their principal values. This may even extend to promoting accretion onto tidal flats. However, the experience from exporting and planting *Spartina* spp. suggests that inappropriate promotion of accreting saltmarsh can cause problems as detailed in Chapter 9. This, together with the arguments associated with combating 'saltmarsh squeeze' (Section 3.5.2.), provide some of the reasons for undertaking coastal realignment (Chapter 6). Using examples from different parts of the world, Section 10.4 describes how re-creating saltmarshes helps improve sea defence in a cost-effective way, as well as improving landscape and nature conservation values. The key areas considered here are:

- Southern North Sea;
- Mediterranean;
- North America.

## 10.4 Southern North Sea

Saltmarshes are extensive around the shores of the southern North Sea particularly in the estuaries of south-east England, the Delta and the Wadden Sea (Figure 5). Losses due to human enclosure are considerable (Section 2.4). This has led to 'saltmarsh squeeze' a combination of loss of habitat through enclosure coupled with erosion due to sea-level rise and increased storminess (Section 3.5.2). The losses observed in south-east England led to the promotion of managed realignment as a cost-effective solution to the twin problems of biodiversity loss and the needs of flood management (Morris et al. 2004). The resulting attempts to restore former enclosed tidal saltmarsh by managed realignment and other restoration techniques have developed apace.

Since 1991, in south-east England there have been 11 examples of complete or partial removal of a seawall allowing the tide to reinvade the land. At a further 4 sites there is some form of regulation of the tidal regime using sluices or one-way valves inserted in the embankment to allow control of tidal levels. From the first experimental managed realignment in the Blackwater Estuary to combat erosion (Section 3.5.1), to the scheme at Freiston, which links flood protection with nature conservation (Section 3.5.3) and in the Humber Estuary (Winn et al. 2003) the schemes have grown progressively larger. The Northey Island experiment amounted to only 0.8 ha whereas on the Humber Estuary, the Alkborough scheme completed in 2004 was 400 ha and a further 1,000 ha is planned later (Wolters et al. 2005). Due to the sinking of the land and the rise in sea level, pressure will continue on the sea defences throughout the area. The response will determine the extent to which wildlife values or human uses prevail.

## 10.4.1 Will it All Come Out in the Wash?

Chapter 3 provides a view of the evolution of thinking on the values associated with saltmarshes. This view rests on an historical appreciation of the functioning of saltmarshes in relation to the natural forcing factors of time, tides and sea-level change. It also provides information on the way human intervention has claimed large areas of land from the sea, causing a 'coastal squeeze'. Looking back at the evolution of the area provides a picture of a once extensive coastal margin, which moved landward or seaward depending on the rate and direction of change in sea level.

Up to about 4,400 years ago, the Fenland Basin was flooded and showed a wholesale landward movement of the sea. During this period, sediment began to fill up the basin, which eventually led to a reversal of the direction of movement such that the shoreline (saltmarsh and tidal flats) began to move seawards again approximately 4,000 years ago. Localised at first, the movement became more widespread until about 3,000 years ago, when it changed once again and became one of a landward transgression by the sea. A reduction in suitable sediment appears to have caused this change in direction (Brew & Williams 2002). The seaward progression of the land, from about 2,000 years ago to the present, is probably as much to do with human activities associated with saltmarsh enclosure (Section 2.3.1) as with natural processes.

The recognition that continued enclosure would result in a loss of intertidal land, coupled with changing economic circumstances, led to the cessation of land claim (Section 3.5). It seems likely that with sea levels rising because of global warming, the next phase in the 'natural' cycle would be a landward progression of the sea. This will inevitably put pressure on the sea defences erected over several centuries. Many of the defences were earth banks, with a crest height insufficient to accommodate the predicted rise in sea level.

The response in past decades was to strengthen these defences by building sea walls that were higher and wider (Figure 69).

**Figure 69** Raising the Wash banks in September 1980. Part of the last major rebuilding programme involving excavating mature saltmarsh to provide material for the 'improved' sea bank

Since this programme of repair, no further enclosures have taken place and in a few places, there has been a reversal of the process. It is ironic that the last enclosure on the Wash in 1985 at Freiston, should be the first managed realignment, designed to take account of the risk of flooding and to improve the nature conservation aspects of the site (Section 3.5.3). It is unclear as to the extent of further realignments around the seaward margins of the Fenland Basin. However, given that most of the land is below the normal High Water Mark of Spring Tides, with some areas below mean sea level (Figure 70), the threat from sea-level rise is considerable. The prognosis under various global warming predictions into the 2050s and beyond, suggests inundation of considerable areas of East Anglia, because of coastal and river flooding (Nicholls & Wilson 2001). A storm surge such as occurred in 1953, given the higher tide levels, has the potential to have even greater impact, despite the improved sea defences.

Within the former peatlands around the inland edge of the fen, a 'Great Fen Project' (see http://www.greatfen.org.uk/) is unfolding. Here, the local Wildlife Trust and others are promoting the extension of wetlands around two existing nature reserves, in an attempt to re-create a self-sustaining reserve through the reclamation of arable farmland. This project is the largest and most ambitious of a number of wetland restoration schemes taking place around the landward margins of the Fenland Basin.

It is unlikely that this initiative will link up with the existing and proposed managed realignments on the current coastal margin of the Wash, because of the high quality agricultural land and the number of settlements that lie in between. The continued protection of the land from flooding by the sea will remain a priority. However, if some of the extreme predictions of global warming and the associated rise in sea level are realised, then protecting this land will become more and more

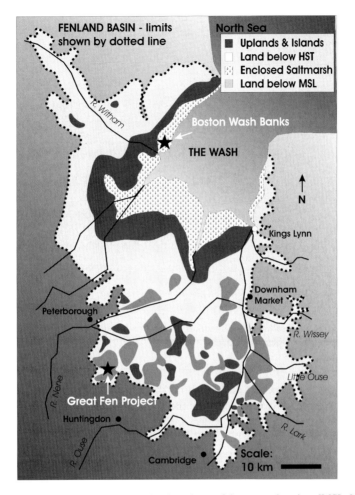

**Figure 70** Land in the Fenland basin, sea levels and two of the restoration sites. [MSL, Mean Sea Level; HST, level of normal High Spring Tides]

difficult and costly. Projecting into the future, we may yet see extensive areas of farmland, currently lying below mean sea level, with their attendant wildlife, stretching from the former landward extremities of tidal influence, to the sea.

## 10.4.2  Realignment in Belgium

In the Yzer River mouth, Belgium attempts to restore intertidal land met with some success. During the twentieth century on the eastern bank of the river, building a military harbour, together with dumping of dredged material from the harbour,

**Figure 71** Developing saltmarsh on mudflats where a naval base once stood, within the Yzer-rivermouth restoration, Belgium coast, 2005

resulted in the destruction of all the saltmarshes and sand dunes in the area. A plan to restore this site, drawn up in 1996 and put into practice from 1999 onwards, formed part of an 'Integrated Coastal Conservation Initiative', funded under A LIFE Nature project (LIFE96 NAT/B/003032). The works involved removing the military buildings and harbour and the removal of the 'protecting' sea walls, a form of managed realignment (Deboeuf & Herrier 2002). This phase of the work was only partially successful in recreating saltmarsh, as the absence of any protection (the sea walls were removed) resulted in erosion of the tidal flats within the site. The final phase of the work involved excavating the dredged material, leaving some of the walls in situ to provide protection for the developing tidal flats. The resulting habitats included the development of pioneer *Salicornia* spp. and *Suaeda maritima*, after only one year (Herrier et al. 2005; Figure 71).

## *10.4.3 The Wadden Sea*

In the Dutch Wadden Sea, the main consideration is habitat creation for the enhancement of nature conservation and landscape values. There have been four deliberate breaches of sea walls ranging between 23 ha on the island of Terschelling to 135 ha in Friesland, as well as two examples where regulated tidal exchange has been used (Wolters et al. 2005). There appears to be little information on managed realignment schemes in the Danish part of the Wadden Sea. Stricter planning controls within the coastal zone, coupled with a more or less stable relative sea level make the problems of flooding and erosion less politically sensitive (Anon 2002).

In Germany, there are two managed realignment sites in the southern North Sea, both in the Wadden Sea; these involve an 80 ha breach in 1994 and a further 280 ha in 1995 both using regulated tidal exchange. Further breaches are planned (Wolters et al. 2005). Also in the German Wadden Sea, embankment of 3,342 ha of Nordstrand Bay (within the Schleswig-Holstein Wadden Sea) took place in 1987 to reduce coastal erosion and shorten the sea defences. The new polder (Beltringharder Koog) included a salt-water lagoon of 846 ha, designed to help compensate for the substantial loss of tidal flats and saltmarshes, within an important wildlife area. For the first few years, there was no tidal inundation and the saltmarsh vegetation soon gave way to grasses such as Rough Meadow-grass (*Poa trivialis*) and Yorkshire-fog (*Holcus lanatus*), not tolerant of sea water (Wolfram et al. 1998).

From 1990, two sluices linked the lagoon to the sea. In 1994, an improvement to the sluices allowed control of flows during natural tidal cycles of the Wadden Sea. This resulted in a tenfold reduction in the tidal range when compared with the outside (0.2–0.4 m tides), which reduced scour near the sluices and did not cover the higher levels within the lagoon. To achieve this, opening the sluices about twice a month to simulate storm-flood conditions, raised the levels to 0.8 m. Studies of the vegetation succession revealed that after 10 years, the semi-natural tidal regime in the lagoon did create saltmarsh vegetation, which became a refuge for rare halophytes (Wolfram et al. 1998). However, it did not compensate for the loss of former Wadden Sea habitats, in particular for wintering waterfowl, although it provided an important roosting area for birds during high tides (Hötker 1997).

## 10.5 Restoration in North America

From the time of the early settlers during the 1700s, wetlands were disease-ridden swampy lands of little use to frontier survival. The Federal Government encouraged land drainage and wetland destruction through a variety of legislative and policy instruments. By the 1960s, most political, financial, and institutional incentives to drain or destroy wetlands were in place. However, since the 1970s increasing awareness that wetlands are valuable areas has helped to reverse Federal and State policies (Dahl & Allord 1997). The 1972 Coastal Zone Management Act and the 1982 Coastal Barriers Resources Act provided for the protection of coastal wetlands. The 'Restore America's Estuaries' campaign (see http://www.estuaries.org/) goes further by setting the framework for restoration of wetlands, including saltmarshes. As the leader in national efforts to protect and conserve the nation's estuaries, it is working to 'restore 1 million acres of estuarine habitat by the year 2010'. In the face of global warming and the associated sea-level rise, this may be a much too cautious approach. Titus (1991) suggests that the USA could afford to lose an 'area the size of Massachusetts' as part of an effort to cope with sea-level rise.

## 10.5.1 The State of Louisiana and the Mississippi Delta

The growth of the Mississippi Delta results from periods of major sediment transport to the system, occurring over several thousand years. In the last few decades, there has been a reversal of this wetland expansion because of human activity. This amongst other things has greatly reduced the input of sediment (Day et al. 1995), such that today regional subsidence in the Delta is about 10 mm per year. As a result, the State of Louisiana has lost up to 40 square miles of marsh a year for several decades, representing approximately 80% of the nation's annual coastal wetland loss. At this rate, by the year 2040, an additional 800,000 acres of wetlands could disappear with the Louisiana shoreline advancing inland by as much as 33 miles in some areas.

These rates of loss prompted Congress to pass the Coastal Wetlands Planning, Protection and Restoration Act (CWPPRA) in 1990, which provides funds for planning and implementing projects. It aims to acquire, restore, manage, or enhance coastal wetlands to help restore the efficiency with which the shoreline adjusts to wave and tidal energy, aiding coastal defence and flood alleviation. In this context, there is recognition that saltmarshes, as part of the coastal wetlands habitat, help provide a coastal protection function.

Approximately $50 million is spent annually on the restoration projects. The official Coastal Protection and Restoration Authority (CPRA) web site (see http://lacoast.gov/) lists 159 projects throughout the State of Louisiana. Between 1991 and 2006, there were 78 projects completed, in the course of construction or approved. Of these, techniques helping to restore saltmarshes include vegetation planting, hydrological restoration and marsh creation. A summary report provides information on the past, present and future status of the Task Force (Louisiana Coastal Wetlands Conservation and Restoration Task Force 2006).

In the wake of Hurricane Katrina (2005), Congress directed the Corps of Engineers, New Orleans District, together with the State of Louisiana, to develop the 'Louisiana Coastal Protection and Restoration (LACPR) Project'. The project has to 'identify, describe and propose a full range of flood control, coastal restoration and hurricane protection measures for South Louisiana'. It is a sign of these more enlightened times that the value of Louisiana's coastal wetlands, including their saltmarshes are in the front line of defence. To quote from the report, 'The people of coastal Louisiana are engaged in a battle against the encroaching Gulf of Mexico. A tenet of efforts to restore and sustain coastal ecosystems dictates that *risk reduction measures not destroy these resources*. As such, plans to restore coastal features as natural lines of defence are an integral part of an overall storm risk reduction and survival plan for Louisiana' (US Army Corps of Engineers 2006).

The events in New Orleans caused by Hurricane Katrina, have also given strength to the consideration of other major projects, one of which involves restoring Louisiana's vanishing wetlands by deconstructing the levee system that controls the Mississippi River. 'Time to move the Mississippi', said a group of experts in coastal management at a meeting in April 2006 reports Cornelia

Dean of the New York Times. Discussion centres on instigating a major river diversion, below New Orleans, that would allow the silt-laden river to flood large areas of the state's sediment-starved marshes. Get more information from the campaign web site to save coastal Louisiana, 'Restore America's Wetland' (see http://americaswetland.com/).

## 10.5.2 San Francisco Bay

The San Francisco Estuary Project (see http://sfep.abag.ca.gov/) is one of more than 20 National Estuary Programme projects. It includes a specific wetlands restoration programme for the San Francisco Bay area (see http://www.sfwetlands.ca.gov/). The history of restoration has moved from a few small scale areas in the 1960s to the thousands of ha envisaged today (San Francisco Estuary Project 2005). The South Bay Salt Pond Restoration Project, for example, includes the restoration of commercial salt production areas in the north of the Bay, which has expanded to over 6,000 ha in the south. Reopening many of the evaporation ponds to tidal influence will facilitate this change. The remaining ponds will have their water levels managed for the benefit of wildlife.

In the San Joaquin-Sacramento River Delta, much of the area was enclosed and drained for agriculture in the late 1800s and early 1900s. Restoration includes setting back levees from the river and restoring tidal influence to the lower, western sections of the delta (San Francisco Estuary Project 2000).

## 10.5.3 Restoration in Canada

Enclosure and drainage for conversion to agriculture, since European settlement began, has destroyed 65% of saltmarshes in the upper Bay of Fundy, Canada. Other activities that exclude the tide resulted in further losses. Attempts to restore these lost areas follow some of the methods used in north-west Europe. Thus, breaching sea walls, reintroducing tidal flows, or simply enlarging culverts and plugging drainage ditches occur. Details of these approaches and projects can be found on 'Environment Canada' web site (see http://www.atl.ec.gc.ca/wildlife/salt_marsh/toc_e.html), which is specifically devoted to saltmarsh conservation and restoration.

## 10.6 The Wider Role – Management and Restoration

The book shows how it is possible to change the direction of development in saltmarshes in such a way as to alter the states and values of the habitat. Justifying intervention on nature reserves, by way of grazing management is relatively easy, although there may be a debate as to the level of intervention and the precise regime appropriate to a given site (Chapter 8). Suggesting that further enclosure, development or other forms of habitat loss should cease is less easy. However, the examples of saltmarsh restoration described above show an increasing acceptance in some places that saltmarshes, either in their own right or as part of a wider coastal wetland, are valuable assets. Despite this, there is reluctance in many areas to adopt anything other than small-scale 'experimental' projects.

### 10.6.1 Approaches to Restoration in the Wadden Sea, the Netherlands and Germany

The situation in the Wadden Sea provides a general overview of the problems associated with allowing the sea to 'reclaim' extensive areas of former tidal land. The whole of the area is of high nature conservation value and has important fishery, recreational and other economic values. The three countries concerned with the area, namely The Netherlands, Denmark and Germany, have been working together since 1978 to protect and conserve the area (see the Wadden Sea web site http://www.waddensea-secretariat.org/). With most of the tidal land under various nature conservation designations, long-term sustainability of the socio-economic and conservation integrity of the area, as well as securing the protection of the hinterland from flooding by the sea are primary objectives (Stade Declaration 1998).

So far, there is acceptance that reducing grazing pressure and drainage on saltmarshes is important for nature conservation purposes. This change also recognises the landscape and cultural value of saltmarsh. Although the primary aim is to maintain the habitat, it also contributes to sea defence.

In Lower Saxony, Germany, the maintenance of 'summerdikes', formerly important in allowing extended agricultural use of saltmarsh (for grazing), is now weighed against their value for nature conservation or sea defence (Ahlhorn & Kunz 2002). Where possible the overall aim is to maintain and extend saltmarsh and enhance their natural development (Bakker et al. 2005). However, this may not be enough. The rigid line of sea walls prevents the dynamic development of intertidal habitats, including saltmarshes, such that there is a diminution in the ability of the Wadden Sea to accommodate a rise in sea level, increased storminess and tidal surges. Thus, in addition to the restoration of 'natural' saltmarshes and other forms of 'soft' sea defences, future prospects may require approaches that are more radical. These could include moving structures such as harbours offshore, raising houses above high-tide levels, or creating areas for floodwater storage (Reise 2005).

In both the Netherlands and Germany (Niedersachsen) some limited realignment has taken place for nature conservation purposes (see above). These include approximately 120ha from the naturally breached Peazemerlannen polder in the Wadden Sea (Bakker et al. 2002). However, Schleswig-Holstein, Germany includes coastal flood-prone lowlands, assets at risk and over a third of a million people. There is also a long history of coastal defence and good quality sea walls, making abandoning any of these areas, through re-integration with the sea a difficult choice. Whilst there is recognition that saltmarsh forms part of the sea defence along the coast and has nature conservation values, the recently adopted 'coastal defence master plan' specifically says:

> "The relocation or abandonment of sea walls remain exceptions (Hofstede 2004)."

### 10.6.2 Depoldering, the Delta Region of the Netherlands

In order to ensure the accessibility of the harbour of Antwerp (Belgium), through the Westerschelde estuary (the Netherlands), the shipping channel is regularly maintained and has been deepened over the last 50 years in order to allow bigger ships to pass through it. This has resulted in coastal erosion on both sides of the estuary with a loss of tidal sand and mud flats and saltmarshes. (These habitats form part of a Special Protection Area, designated under the European Union Birds Directive). Under the terms of the Habitats and Species Directive, the Dutch Ministry of Transport, Public Works and Water Management proposed compensation for this loss through a strategy of managed realignment in 1998. The works proposed included cutting dykes into several polders (themselves derived from former tidal land) allowing them to become tidal again.

Fearing that a repeat of the 1953 floods, killing many people and thousands of cattle might occur, local people resisted the proposal. In 2000, the local authorities rejected the 'depoldering' plan. The government withdrew the proposal soon after. It is clear that in the Netherlands the extent of flooding and loss of life in 1953 remains a powerful force, which militates against allowing the deliberate breaching of existing sea defences in the Delta Region. Several other breaches proposed for the Westerschelde in 1995 were not been carried out (Wolters et al. 2005). By contrast, an unplanned breach in a brackish part of the Scheldt estuary occurred during a severe storm in 1990, returning tidal influence to about 100 ha of the Sieperda polder. There was no attempt to repair the breach and in 10 years, the former polder changed into a brackish tidal marsh with a wide range of typical plants and animals (Eertman et al. 2002).

### 10.6.3 The Situation in the UK, Winning Hearts and Minds

The identification of 'Coastal Cells' provides the basis for defining the boundaries of management units (Figure 72). These represent a series of interlinked systems where sediment movement by waves and currents defines sediment transport cells. The cells and subcells identified for England and Wales and Scotland comprise an arrangement of:

- Sediment sources (e.g. eroding cliffs, river, sea bed);
- Areas where sediment is moved by coastal processes;
- Sediment stores or sinks (e.g. beaches, estuaries and offshore banks).

**Figure 72** Shoreline management units for the coast of Great Britain

The Government has promoted the formation of voluntary coastal defence groups around these coastal cells, made up of maritime district authorities and other bodies with coastal defence responsibilities. Shoreline Management Plans help provide, at a large-scale, an assessment of the risks associated with changes in the coastal environment. They aim to reduce the risks to socio-economic infrastructure, as well as the historic and natural environment, from flooding and coastal erosion.

Amongst the policy options is managed realignment, already described. Schemes such as Freiston (Section 3.5.3; Figure 24) and Alkborough (Section 6.4.2; Figure 52) in particular, bring together flood alleviation, with other environmental benefits. Amongst these benefits is the re-creation of saltmarsh. Detailed information on the development of these policies is available on the UK Department of Environment, Food and Rural Affairs (2001); see web site (http://www.defra.gov.uk/environ/fcd/policy/smp.htm).

Even in the UK where there are now 59 managed realignment schemes (Wolters et al. 2005) only three are over 100 ha in extent. Each proposed scheme requires a detailed (and expensive) feasibility study and includes extensive local consultation amongst stakeholders. The Alkborough project in the Humber Estuary, in particular, provides an example of this inclusive approach. The review of shoreline management in the Humber Estuary, in 2001 included proposals to realign an area at Alkborough.

'Selling' the proposal involved extensive consultation with local stakeholders about the project's aims and objectives. Specifically it should offer:

1. A sustainable solution;
2. Reduce flood risk;
3. Create new wildlife habitats;
4. Develop new economic opportunities;
5. Provide new recreational opportunities.

The arguments included the effects of climate change, which would increase high-tide levels in the Humber Estuary. As a result, leaving the defences as they are would put the homes of 300,000 people living in the area at risk from flooding. Overall, the scheme also provided significant environmental enhancement, including restoring part of the functioning of the estuary. The key to its successful completion lay partly in providing greater certainty about the sustainability of nature conservation, economic and recreational opportunities in the face of rising sea levels. The extensive consultation was instrumental in gaining acceptance of a scheme that appeared to involve the loss of 440 ha of prime agricultural land.

This enabled the Government Minister (Elliot Morley) in 2005 to claim that the UK had taken 'a new direction in flood risk management'. 'Alkborough is a fine example of a sustainable approach to reducing flood risk by working with the forces of nature. Such long-term solutions are essential if we are to protect the lives and homes of people who live and work in the area, as well as the many businesses that are based around the estuary.'

Finally launched in September 2006, the scheme cost £10.2 million. The money came from a range of governmental organisations and the sources included the Department of Environment, Food and Rural Affairs, the Regional Development Agency, English Nature, the Heritage Lottery Fund and the European Union (via the Interreg programme). The scheme will re-create an intertidal area with saltmarsh and mudflats, providing a focus for education and access for local communities.

### *10.6.4 The Mediterranean, Sediments and Deltas*

In the Mediterranean, recognition of the issues surrounding global warming and sea-level rise does not appear to be a major concern. However, a Global Vulnerability Analysis suggests that the Mediterranean coast is generally more prone to sea-level rise than many other parts of the world, both in relation to economic and environmental costs (Nichols & Hoozemans 2005). The larger saltmarshes occur in the Po Delta and the Venice lagoon, Italy, the Ebro Delta, Spain, the Rhône Delta (Camargue), southern France and the in the Axios Rivers, Greece. A combination of factors including deltaic subsidence, reduced sediment supply (through river damming) and enclosure of tidal areas, including saltmarshes, contribute to the high rates of relative sea-level rise occurring in these areas. The extent of loss due to human intervention, the inability of the habitats to accrete in this microtidal sea and the restriction on landward movement due to artificial defences, prevent them keeping pace with sea-level rise. Most of the deltas have rates of relative sea-level rise, which could reach 10mm per year over the next century. These are comparable to those experienced in the Mississippi Delta (Day et al. 1995). Some individual examples highlight the issues.

## 10.6.5 The Ebro Delta

The Ebro Delta saltmarshes include transitional reedbeds (*Phragmites australis*) and a *Salicornia* community (*Arthrocnemum fruticosum*). As elsewhere these and other tidal habitats have been enclosed and developed, in this case mainly for rice production (57% of the total area). Damming the Ebro River has resulted in reduction in river flows and a major reduction in the sediment supply (Ibàñez et al. 1996). With an average rate of relative sea-level rise of 3mm per year it is only in the wetlands around the river mouth that accretion rates (4mm per year) exceed this figure. As a result, there is wetland loss, coastal erosion and saltwater intrusion. The reversal of these trends is essential if the wetland habitats and rice cultivation areas can grow vertically and keep pace with the relative sea-level rise. The solution to this lies in reinstating the flow of sediments from eroded land in the hinterland, through the river dams where it is currently trapped (Ibàñez et al. 1997).

A review of the conservation issues as they affect the area took place in September 2000 (Viñals et al. 2001). Amongst the many problems identified was the lack of sediment delivery (95% reduction) mainly due to the river dams. The resulting erosion and saline intrusion threatened many aspects of the wildlife of the area. The Spanish National Hydrological Plan would further exacerbate these problems if, following approval by Congress in 2001, they were to be implemented. This plan was an attempt by the Spanish Government to assure constant water supply all over Spain. In the Ebro Delta, the proposal involved the transfer of the 'surplus flow' in the Ebro watershed to other, southern river basins.

The issue was highly contentious. The Worldwide Fund for Nature (WWF) and others campaigned against what they saw as an 'immensely dangerous project'. The Spanish government announced an alternative plan in June 2004 to replace the Ebro Transfer project, although it is not clear if this is the end of the threat. The RiverNet web site provides a comprehensive description of the scheme and the outcome of the opposition (see http://www.rivernet.org/Iberian/planhydro.htm). Although it appears that this last threat is no longer extant, the site will continue to suffer from the effects of relative sea-level rise. Its long-term sustainability both for wildlife (including the saltmarshes) and rice cultivation depends on the reinstatement of the sediment flows to the Delta.

## 10.6.6  The Venice Lagoon

In the Venice lagoon, where sea-level rise is 2.5 mm per year, saltmarshes in the central area erode due to sea-level rise, and although vertical accretion takes place, this is at the expense of sediment derived from the eroding saltmarsh edge (Day et al. 1998). Climate change and sea-level rise are likely to lead to the complete disappearance of these saltmarshes over the next 30–50 years, if natural and artificial adaptive responses are not implemented (Brochier & Ramieri 2001).

A plan for the restoration of the lagoon by the Consorzio Venezia Nuova (CVN) includes reinstatement of sediments lost to the system. Accompanying this is the construction or repair of physical structures as well as natural features such as channels, shallows, saltmarshes and mudflats. The plan to reverse the environmental deterioration also aims to improve sediment and water quality. Today an extensive and ambitious series of restoration measures are underway. These include 'Environmental defence: protection and reconstruction of mudflats and saltmarshes habitat and structure'. Methods include the use of dredged material to aid the restoration of saltmarsh and mudflats. For more information, see (http://www.salve.it/uk/default.htm). During the process of restoration the discovery of a fourteenth century galley sunken in a previous attempt at coastal protection (Merali 2002), suggests that the problems of erosion and flooding are not new. It is interesting to speculate what future historians will make of the sunken Thames Barges off the Essex coast (Section 6.2.6; Figure 47)!

## 10.7  The Future

It appears that the case for saltmarsh restoration, as part of wider wetland re-creation and creation, is accepted policy in many parts of the world. Saltmarshes are important components of protected areas and nature reserves. They are no longer looked at just for their value in the creation of agricultural land or 'wastelands' fit only for dumping rubbish or enclosure for industrial and housing development. Rather their values are much more wide ranging. They are important nature conservation habitats both in their own right and as components of interrelated coastal ecosystems. They help provide protection from coastal flooding and erosion. They add to the overall landscape and recreational experience.

Whether the trend towards abandonment of land to the sea becomes more widely accepted as an effective means of sea defence, especially in areas of rising sea level, is unclear. The evidence from around the world, not least in the Mississippi Delta, suggests there is an increasing realisation, if current trends continue some

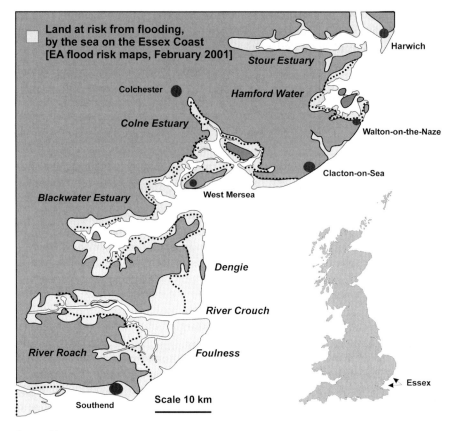

**Figure 73** Coastal Essex, Great Britain. Land at risk from flooding by the sea, Environment Agency Flood Risk Maps (shown in light grey) in relation to the area flooded during the 1953 storm surge, which took place in January/February. The dotted line shows the approximate limit of flooding in 1953, derived from a map in Grieve (1959)

areas will be lost or rendered uninhabitable. It is almost certainly better to plan for this by strategic retreat, than continue the battle with the sea! Many of the realignment schemes in the UK are concentrated in south-east England. The area of land lying on the coast of Essex that has the potential for flooding when another storm surge such as 1953 occurs, could be viewed as land of potential value for restoring saltmarshes (Figure 73).

## 10.7.1 A European Initiative

At a European level under the European Union Community Initiative Programme Interreg IIIB for the North Sea, includes a project which aims to create multifunctional flood management schemes with a more gradual transition from sea to land. The project 'ComCoast' runs from April 2004 to December 2007 at a cost of €5.8 million. It embodies some of the principles already incorporated into the realignment schemes such as Alkborough and Freiston. It focuses on embanked coastal areas and aims to provide 'economic, sustainable alternatives to the traditional single line flood defence'. Further information on the project is available from the web site (see http://www.comcoast.org/).

If this is the adopted course of action, there are fundamental questions for coastal defence authorities as well as nature conservation organisations. Not least of these, is the way in which trade-offs will need to be made. For the coastal defence authorities land will not always be sacrosanct! It is unlikely we will reach the 'Water-world' depicted in the film by Kevin Costner. However, given the area of agricultural and other land potentially at risk from flooding in low-lying areas, it is inevitable that re-integration with the sea will take place over larger areas, in order to secure cost-effective sea defences.

## *10.7.2 A Historical Perspective*

The history of human involvement on the coast has moved from one where living **by** the sea meant living **with** the sea. The earlier settlers found the coastal margins, particularly those of coastal embayments, some of the richest areas to find food and shelter. During the early stages of the Holocene, when sea levels were much lower than today there is evidence of many settlements below the present low water mark. In Denmark, for example, there is a well-preserved settlement of Mesolithic age at the Tybrind Vig site. The site is 300 metres from the shore and 3 metres below the surface, and includes well-preserved artefacts from the Ertebølle Culture (Malm 1995). In the Mediterranean, there are Neolithic remains between 0.5–5 m below present day sea level dating from 7,100–6,300 years ago. Other remains dating back to 8,180–7,550 years ago occur at Atlit-Yam, situated some 200–400 m from the coast of Israel, at a depth of 8–12 m (Information from the Israel Antiquities Authority web site, see http://www.antiquities.org.il/home_eng.asp). Settlements were, however, at the mercy of the sea-level change, tides and storms. These and many other sites, along with the forests that occurred there have long since been overwhelmed by rising seas.

As sea levels began to stabilise around 7,000 years ago, coastal human settlements grew. Coastal habitats, especially saltmarshes, provided pasture and hay for domestic animals, in Europe probably before Roman Times. In heavily wooded areas, such as the early settlements in America, the value of this crop as fuel for horses may have been particularly significant. As engineering techniques became more sophisticated and machinery more powerful, the ability to alter the coastal margin and 'control' the sea became easier. This not only led to the expansion of agriculture at the expense of saltmarshes, but also to the enclosure of tidal lands for ports, harbours, industrial, other infrastructure and housing development (Chapter 2). Until the middle of the twentieth century, the implications of these losses for the environment were secondary to the economic benefit that accrued.

### 10.7.3  *A New Perspective on Saltmarsh Conservation?*

Even in the early 1990s notions that we should (or could) give up land to the sea were an anathema to many. In the field of engineering, for example, the River Engineering Section of the United Kingdom Institution of Water and Environmental Management (IWEM) on 1 February 1991 politely received a review of the relationship between sea defence and nature conservation. The presentation argued that we may have "devoted too much effort to the 'battle with the sea'". and that adopting a more flexible approach to coastal defence 'might provide opportunities for positive nature conservation action and more cost effective sea defences.' Compliance with a request for the author to submit the paper to the IWEM Journal resulted in its rejection.

The editorial panel in assessing the paper 'agreed that (it) puts forward a fresh and interesting idea, but did not convince them that we could safely leave coastal protection to natural processes, except in a few fallible situations'. In fact, the author suggested a policy involving developing a strategy where the allocation of resources was to "areas where the needs of sea defence and coast protection were paramount, provides opportunities for 'soft' engineering and increases the size of the zone in which the natural sea defences can operate" (Doody 1992).

Advocating this more flexible approach, and with it a recognition of the role of natural habitats in coastal defence, is not new. In this context, it is instructive to read part of the final report of the UK Royal Commission on Coastal Erosion, 1911:

> "The rate of erosion varies with the geological formation of the coast, but is most marked along the east and south coasts of England. On the other hand there have been considerable gains, particularly in the mouth of the Humber and the Wash. Natural protection is afforded to the coast by the foreshore and beach material produced by erosion and it is essential that such material should not be removed. In some places erosion has been aggravated by the erection of defences of the wrong type. The Central Authority, aided by scientific experts, should make systematic observations of coastal changes. In late years the gains had generally outweighed losses, but this took no account of the value of the property, especially as many accretions were below high-water level. There had been some serious losses calling for effective measures of prevention. The cost of protecting purely agricultural land will usually exceed the value of the land and such works should be undertaken only when they preserve a considerable area of low-lying land."

Chapter 3 suggests that in the short space of 30 years, a reversal of the notion that we could continue to 'win' land from the sea with few environmental consequences, has taken place. Initially, the drive towards this in the UK was the recognition of nature conservation losses, including those of saltmarsh. Over 1,000 ha of saltmarsh (one quarter of the resource) suffered erosion on the Essex coast between 1973 and 1998 (Cooper et al. 2001). This was in addition to the approximately 3,000 ha lost to enclosure in historical times. Other calculations suggest that sea-level rise will result in the further loss of 8,000–10,000 ha of mudflats and hence potential areas for saltmarsh development, in Britain between 1992 and 2012 (Pye & French 1993). These actual and predicted losses are unacceptable.

Against this background, maintaining the existing resource in the UK, or elsewhere in the world is not enough. This derives not only because of the implications for nature conservation but also for its many other values. Not least amongst these, are those associated with sea defence and the functioning of the tidal ecosystem. The reversal in policy, which allows breaching of the existing sea walls, represents a major change in the approach to sea defence. Amongst the reasons for this change in attitude are the following, realignment sites:

1. Provide sustainable and effective flood and coastal defence;
2. Are essential for a long term strategy of coping with sea-level rise;
3. Provide environmental benefits in terms of habitat creation;
4. Support the Habitats Regulations (by providing a means of compensating for intertidal habitats lost elsewhere through coastal squeeze);
5. Reduce costs of flood and coastal defence;
6. Are better than dealing with an accidental breach;
7. Are a low cost means of re-creating natural habitats (Anon 2002).

In this wider context, by adopting a more flexible approach to habitat and species protection, it is possible to see new and more innovative solutions to the conservation of saltmarshes and coastal systems more generally. These build on:

1. A better understanding of the geomorphology of the coast and the role of saltmarshes in the functioning of the estuary and the relationship with the catchment **and** near shore marine environment;
2. Recognition of the value of saltmarsh, not just for wildlife but also for other environmental and socio-economic interests;
3. An acceptance that change is a natural part of the saltmarsh development, particularly in response to changes in sea level;
4. Restoration of saltmarshes, which not only re-creates inherent values, but also helps heal degraded coastal systems.

This book has attempted to identify the common ground between those concerned with nature conservation and other forms of coastal use in relation to managing saltmarshes. From both perspectives, it would appear that protecting saltmarshes in the traditional sense is, in many areas, ultimately doomed to failure. With the existing rise in sea level, even with the curtailment of destructive land-based uses, they will continue to be lost. However, accepting that the natural ability of the coastline and coastal habitats to adapt to change should be encouraged is not easy.

Innate prejudices and fears about the local impact on lives and property militate against allowing the sea to invade the land, either through neglect or as a deliberate policy. From a nature conservation perspective, the approach is no less easy to accept especially when it involves the sacrifice of one treasured habitat, such as coastal grazing marsh, for another.

## 10.8 What Does the Future Hold?

Saltmarshes will continue to form a link between the land and the sea and a self-regulating coastal defence. However, as sea-level rises and sediment deficits continue, a question arises as to their long-term viability. As already indicated the prognosis is not good (Nicholls et al. 1999; Nicholls & Hoozemans 2005). Saltmarshes are amongst the first habitats affected by rising sea levels due to global warming. We appear to have two choices:

1. Continue to believe that we can 'hold back the sea' in most instances and build bigger and 'better' sea defences. This will inevitably result in continuing saltmarsh losses as 'coastal squeeze' takes place;
2. Recognise that saltmarshes play an important and valuable role in many aspects of human activity. Making space for them to do what comes naturally, and move in response to changing environmental conditions, may yet help to deliver sustainable economic, social and environment benefits.

In the face of global warming, there is a move to increase the use of renewable forms of energy generation, including tidal energy. However, these are not benign and saltmarshes are likely to be one of the more affected habitats (Clark 2006). This poses a final dilemma for the saltmarsh specialist, can we accept the loss of this fascinating habitat for the greater good of the world. The answer is an emphatic **no!** Saltmarshes form an integral and significant part of the general development and flexibility of our coastal ecosystems. Without them, the coast would be aesthetically poorer as well as lacking many of the ecosystem service values they support. It is in all our interests to continue to protect and manage them and where necessary restore and re-create them.

# Appendix – English and Latin Names

## Birds

| | |
|---|---|
| Avocet | *Recurvirostra avosetta* |
| Black-winged Stilt | *Himantopus himantopus* |
| Brent Goose | *Branta bernicla* |
| California Black Rail | *Laterallus jamaicensis coturniculus* |
| Chaffinch | *Fringilla coelebs* |
| Coastal Plain Swamp Sparrow | *Melospiza geogiana nigrescens* |
| Dark-bellied Brent Goose | *Branta bernicla bernicla* |
| Eastern Black Rail | *Laterallus jamaicensis jamaicensis* |
| Flamingo | *Phoenicopterus ruber roseus* |
| Greater White-fronted Goose | *Anser albifrons albifrons* |
| Greylag Goose | *Anser anser* |
| Lapwing | *Vanellus vanellus* |
| Lesser Snow Goose | *Chen caerulescens caerulescens* |
| Little Tern | *Sterna albifrons* |
| Little-ringed Plover | *Charadrius dubius* |
| Marsh Wren | *Cistothorus palustris* |
| Meadow Pipit | *Anthus pratensis* |
| Oystercatcher | *Haematopus ostralegus* |
| Pink-footed Goose | *Anser brachyrhynchus* |
| Redpoll | *Carduelis flammea* |
| Redshank | *Tringa totanus* |
| Saltmarsh Savannah Sparrows | *Passerculus sandwichensis* |
| Saltmarsh Sharp-tailed Sparrows | *Ammodramus cauducatus* spp |
| Seaside Sparrows | *Ammodramus maritimus* spp |
| Shelduck | *Tadorna tadorna* |
| Shorelark | *Eremophila alpestris* |
| Skylark | *Alauda arvensis* |
| Snow Bunting | *Plectrophenax nivalis* |
| Song Sparrows | *Melospiza melodia* |
| Wigeon | *Anas Penelope* |

# Plants

| | |
|---|---|
| *Armeria maritima* | Sea Thrift |
| *Arthrocnemum macrostachyum* | Glaucous Glasswort |
| *Aster tripolium* | Sea Aster |
| *Atriplex portulacoides* | Sea-purslane |
| *Atriplex prostrata* | Spear-leaved Orache |
| *Blymus rufus* | Saltmarsh Flat-sedge |
| *Bolboschoenus maritimus* | Sea Club-rush |
| *Distichlis spicata* | Saltgrass |
| *Elytrigia atherica* | Sea Couch |
| *Festuca rubra* | Red Fescue |
| *Holcus lanatus* | Yorkshire-fog |
| *Iris pseudacorus* | Yellow Flag |
| *Juncus gerardii* | Saltmeadow Rush |
| *Juncus roemerianus* | Needlegrass Rush |
| *Limonium bellidifolium* | Matted Sea-lavender |
| *Limonium vulgare* | Common Sea-lavender |
| *Phragmites australis* | Common Reed |
| *Plantago maritima* | Sea Plantain |
| *Poa trivialis* | Rough Meadow-grass |
| *Puccinellia maritima* | Common Saltmarsh-grass |
| *Salicornia dolichostachya* | Long-spiked Glasswort |
| *Salicornia europaea* | Glasswort |
| *Salicornia ramosissima* | Purple Glasswort |
| *Salicornia virginica* | Perennial Pickleweed |
| *Scirpus mariqueter* | China, no English name |
| *Seriphidium maritimum* | Sea Wormwood |
| *Spartina alterniflora* | Smooth Cord-grass |
| *Spartina anglica* | Common Cord-grass |
| *Spartina cynosuroides* | Big Cord-grass |
| *Spartina densiflora* | Dense-flowered Cord-grass |
| *Spartina foliosa* | California Cord-grass |
| *Spartina maritima* | Small Cord-grass |
| *Spartina patens* | Saltmeadow Cord-grass |
| *Suaeda maritima* | Annual Sea-blite |
| *Suaeda vera* | Shrubby Sea-blite |
| *Tamarisk gallica* | Tamarisk |
| *Triglochin maritimum* | Sea Arrowgrass |
| *Typha angustifolia* | Lesser Bulrush (Cattail) |
| *Typha latifolia* | Bulrush |

# Appendix – English and Latin Names

## Other Vertebrates

| | |
|---|---|
| Atlantic Herring | *Clupea harengus* |
| Brown hares | *Lepus europaeus* |
| Carolina Water Snake | *Nerodia sipedon* |
| Diamondback Terrapin | *Malaclemys terrapin* |
| Dover Sole | *Solea soleaI* |
| European Plaice | *Pleuronectes platessa* |
| Florida Saltmarsh Vole | *Microtus pennsylvanicus dukecampbelli* |
| Grey Mullets | e.g. *Liza ramada* |
| Marsh Snake | *Nerodia fasciata* ssp. *Taeniata* |
| Muskrat | *Ondatra zibethicus* |
| Northern Brown Snake | *Storeria dekayi* |
| Nutria | *Myocastor coypus* |
| Saltmarsh Harvest Mouse | *Reithrodontomys raviventris* |
| Saltmarsh Topminnow | *Fundulus jenkinsi* |
| Sea Bass | *Dicentrarchus labrax* |
| Sika deer | *Cervus nippon* |

## Invertebrates

| | |
|---|---|
| Mud Snail | *Hydrobia ulvae* |
| Essex Emerald | *Thetidia smaragdaria* |
| Halophilic spiders | *Arctosa fulvolineata* (nocturnal lycosid) |
| | *Pardosa purbeckensis* (diurnal lycosid) |
| Common Marsh Hopper | *Orchestia grillus* |
| Isopod | *Philoscia vittata* |
| Beachflea or Beach-hopper | *Orchestia gammarellus* |
| Saltmarsh Periwinkle | *Littoraria irrorata* |
| Ragworm | *Nereis diversicolor* |

# Appendix – A Few Useful Web Sites

**Beneficial use of dredged material http://el.erdc.usace.army.mil/dots/budm/budm.cfm**

This site is a collaborative effort between the US Environmental Protection Agency and the US Army Corps of Engineers and describes the general approaches to the beneficial uses of dredged material under three main headings:

1. Agricultural/product uses;
2. Engineered uses;
3. Environmental enhancement.

The site emphasises that a systematic approach to coastal restoration projects is required to ensure successful completion of projects. Five phases are identified as important to the process:

1. Planning;
2. Implementation;
3. Performance assessment;
4. Adaptive management;
5. Dissemination of results.

It provides a large number of examples of successful restoration schemes including many examples from saltmarsh projects. A list of the various studies, organized by habitat type, conducted to improve the understanding and methodology of restoration for saltmarsh cover:

1. Hydrology restoration and tidal channel development;
2. Elevation manipulation;
3. Plant ecology;
4. Fish and invertebrate growth and habitat use.

**The Coastal Practice Network (CoPraNet), INTEREG IIIC http://www.coastalpractice.net/**

The poject is part-financed by the European Union (European Regional Development Fund) within the INTERREG IIIC Programme. The Coastal Practice Network contributes to the establishment of a European network of coastal

Appendix – A Few Useful Web Sites

managers and policy makers. The site provides links to a number of saltmarsh related studies, such as:

- EUROSSAM, European Saltmarshes Modelling
- http://ecobio.univ-rennes1.fr/eurossam/

The EUROSSAM project aims to develop a policy and management tool for the conservation and restoration of saltmarshes.

- ISLED, Impact of Sea Level on Ecosystem Dynamics of saltmarshes
- http://www.labcoastal.co.uk/projectrep1.htm

This work provides a comprehensive insight into the dynamics of the system that counteracts the effects of sea-level rise. The results are integrated into mathematical models to aid decisions on future management and conservation of saltmarshes.

- CoPraNet, Hesketh Outmarsh Managed Realignment. http://www.rspb.org.uk/ourwork/conservation/projects/ribble/hesketh.asp

This is a large (170 ha) realignment scheme on the south bank of the Ribble Estuary, northwest England. The aim is to counter the effects of sea-level rise due to climate change, through the restoration of land enclosed for agriculture in the 1980s to saltmarsh.

- EUROSION, Case Study: Elbe Estuary (Germany) download report from http://www.eurosion.org/shoreline/16elbe.html

The study includes work on saltmarshes protected by a variety of techniques with an overall change to soft measures since the 1970s. The effect is difficult to assess, as the marshes have a natural capacity to compensate for the effects of sea-level rise and thus accrete and rise in level.

**National Oceanic and Atmospheric Administration** (NOAA) Coastal Services Center **http://www.csc.noaa.gov/coastal/**

The NOAA Coastal Services Center is an office within the National Oceanic and Atmospheric Administration devoted to serving the nation's state and local coastal resource management programs. The site provides links to a wealth of information on coastal issues including Coastal Ecosystem Restoration, including saltmarsh.

**Environment Agency, England and Wales Saltmarsh Management Manual**
http://www.saltmarshmanagementmanual.co.uk/Index.htm

The *Saltmarsh Management Manual* is a comprehensive guide to saltmarsh management and restoration. It aims to assist coastal and estuarine managers in the identification of the problems or management needs such as maintenance, restoration or enhancement and in the determination of appropriate management responses. It provides links to four main sections, describing:

- Saltmarsh condition
- Causes of change

- Problem definition
- Management techniques

A fifth link provides an appraisal process to help ensure that a good understanding of the saltmarsh condition is developed, so that an informed process of selection can be undertaken.

Steps in appraising and selecting a preferred management option are:

**Step 1** – define management objectives.
**Step 2** – describe the baseline environment.
**Step 3** – identify the cause of change.
**Step 4** – assess the implications of alternative management options for the baseline environment.
**Step 5** – define the extent of any predicted impacts and actions (in environmental, social and financial terms).
**Step 6** – select the option that will best achieve the management objectives (including do-nothing), provide the greatest benefit and the best value for money.
**Step 7** – design and implement a monitoring programme.

# References

Abraham, K.F., Jefferies, R.L. & Alisauskas, R.T., 2005. The dynamics of landscape change and snow geese in mid-continent North America. *Global Change Biology*, **11/6**, 841–855.

Adam, P., 1978. Geographical variation in British saltmarsh vegetation. *Journal of Ecology*, **66**, 339–366.

Adam, P., 1981. Vegetation of British saltmarshes. *New Phytologist*, **88**, 143–196.

Adam, P., 1990. *Saltmarsh Ecology*. Cambridge University Press, Cambridge.

Adam, P., 1995. Saltmarsh. In: *State of the Marine Environment Report for Australia: The Marine Environment - Technical Annex: 1*, L.P. Zann, ed. Source: http://www.deh.gov.au/coasts/publications/somer/annex1/saltmarsh.html.

Adam, P., 2002. Saltmarshes in a time of change. *Environmental Conservation*, **29/1**, 39–61.

Ahlhorn, F & Kunz, H., 2002. The Future of Historically Developed Summerdikes and Polders: A Saltmarsh Use Conflict. In: *Littoral 2002, The Changing Coast*, EUROCOAST ed., 365–374. EUROCOAST / EUCC, Porto - Portugal.

Allan, H.H., 1930. *Spartina townsendii*. A valuable grass for reclamation of tidal mud-flats. *New Zealand Journal of Agriculture* **40**, 189–196.

Allen, H.H., Lazor, R.L. & Webb, J.W., 1990. Stabilization and Development of Marsh Lands. *Beneficial Uses of Dredged Material*. Proceedings of the Gulf Coast Regional Workshop April 26–28, 1988, Galveston, Texas. Technical Report D-90-3, 101–112.

Allen, J., 1997. The Geoarchaeology of land-claim in coastal wetlands: a sketch from Britain and the north-west European Atlantic – North Sea. *Journal Royal Archaeological Institute*, **154**, 1–54.

Allen, J.R.L. & Pye, K., 1992. *Saltmarshes: Morphodynamics, Conservation and Engineering Significance*. Cambridge University Press, Cambridge.

Allen, J.R.L., 2000. Morphodynamics of Holocene saltmarshes: a review sketch from the Atlantic and Southern North Sea coasts of Europe. *Quaternary Science Reviews*, **19/12**, 1155–1231.

An, S., Zhou, C., Wang, Z., Yu, D., Deng, Z. & Chen, L., 2004. *Spartina* in China: Introduction, history, current status and recent research. In: *Spartina 2004, Abstracts from the Third International Conference on Invasive Spartina, San Francisco*, 1. Source: http://www.the-conference.com/2004/spartina/index.php

Anastasiou, C.J. & Brooks, J.R., 2003. Effects of soil Ph. Redox Potential and elevation on survival of *Spartina patens* planted at a west Florida saltmarsh restoration site. *Wetlands*, **23/4**, 845–859.

Andresen, H., Bakker, J.P., Brongers, M., Heydemann, B. & Irmler, U., 1990. Long-term changes of saltmarsh communities by cattle grazing. *Plant Ecology*, **89/2**, 137–148.

Anon, 2002. *Managed Realignment Review, Project Report*. Policy Research Project, FD 2008. Flood and Coastal Defence R&D Programme, DEFRA / Environment Agency.

Anon, 2003. 1953. *U.K. Floods, 50 Year Retrospective*. Risk Management Solutions. Source: http://www.rms.com/NewsPress/1953%20Floods.pdf.

Arcese, P.M., Sogge, K., Marr, A.B., & Patten, M.A., 2002. Song Sparrow (*Melospiza melodia*). In: *The Birds of North America, 704*, eds. A. Poole and F. Gill, The Birds of North America, Inc., Philadelphia, PA. Source: http://bna.birds.cornell.edu/BNA/Login.do.

Atkinson, P.W., Crooks, S., Drewitt, A., Grant, A., Rehfisch, M.M., Sharpe, J. & Tyas, C.J., 2004. Managed realignment in the UK – the first 5 years of colonization by birds. *Ibis*, **146**, 101–110.

Atkinson, P.W., Crooks, S., Grant, A. & Rehfisch, M.M., 2001. *The Success of Creation and Restoration Schemes in Producing Intertidal Habitat for Waterbirds*. English Nature Research Report, **425**, Peterborough. Source: http://www.english-nature.org.uk/pubs/publication/PDF/ENRR425_1.pdf

Aubrey, D.G. & Emery, K.O., 1993. Recent global sea levels and land levels. In: *Climate and Sea Level Change, Observations, Projections and Implications*, R.A. Warrick, E.M. Barrow & T.M.L. Wigley eds., Cambridge University Press, 45–56.

Ayres, D.R. & Lee, A.K.F., 2004. *Spartina densiflora* x *foliosa* hybrids found in San Francisco Bay (Poster). *Third International Conference on Invasive Spartina*. San Francisco, CA.

Ayres, D.R., Smith, D.L., Zaremba, K., Klohr, S. & Strong, D.R., 2004. Spread of Exotic Cordgrasses and Hybrids (*Spartina* sp.) in the Tidal Marshes of San Francisco Bay, California, USA. *Biological Invasions*, **6/2**, 221–231.

Bakker, J.P., 1985. The impact of grazing on plant communities, plant populations and soil conditions on saltmarshes. In: *Ecology of coastal vegetation*, W.G. Beeftink, J. Rozema & A.H.L. Huiskes, eds., *Vegetatio*, **62**, 391–398.

Bakker, J.P., Bos, D. & De Vries, Y., 2003. To graze or not to graze: that is the question. In: *Challenges to the Wadden Sea Area*, W.J. Wolff, K. Essink, A. Kellermann & M.A. Van Leeuwe, eds., Proceedings of the 10th International Scientific Wadden Sea Symposium, 67–88, Ministry of Agriculture, Nature Management and Fisheries, Department of Marine Biology, University of Groningen.

Bakker, J.P., Esselink, P., Dijkema, K.S., van Duin, W.E. & De Jong, D.J., 2002. Restoration of saltmarshes in the Netherlands. *Hydrobiologia*, **478**, 29–51.

Bakker, J.P., P. Esselink, van der Wal, R. & K.S. Dijkema 1997. Options for restoration and management of coastal saltmarshes in Europe. In: *Restoration Ecology and Sustainable Development*, K.M. Urbanska, N.R. Webb & P.J. Edwards, eds., 286–322. Cambridge University Press. Cambridge.

Bakker, JP., Bunje, J., Dijkema, K., Frikke, J., Hecker, N., Kers, B., Körber, P., Kohlus, J. & Stock, M., 2005. Saltmarsh, In: *Wadden Sea Quality Status Report, 2004*, K. Essink, C. Dettmann, H. Farke, K. Laursen, G. Lüerßen, H. Marencic, W. Wiersinga eds., *Wadden Sea Ecosystem*, **19**, 163–179.

Barde, J-P. & Pearce, D.W., 1991. Valuing the Environment, Six Case Studies. Earthscan Publications Ltd.

Bassam, N.E., ed., 1998. *Energy Plant Species: Their Use and Impact on Environment and Development*. James & James/Earthscan, 256 pp.

Baumel, A., Ainouche, M.L., Misset, M.T., Gourret, J-P. & Bayer, R.J., 2003. Genetic evidence for hybridization between the native *Spartina maritima* and the introduced *Spartina alterniflora* (Poaceae) in South-West France: *Spartina × neyrautii* re-examined. *Plant Systematics and Evolution*, **237/1–2**, 87–97.

Bazely, D.R. & Jefferies, R.L., 1986. Changes in the Composition and Standing Crop of Salt-Marsh Communities in Response to the Removal of a Grazer. *Journal of Ecology*, **74/3**, 693–706.

Beeftink, W.G., 1977a. Salt-marshes. In: *The Coastline*, R.S.K. Barnes, ed., 93–122. John Wiley & Sons, Chichester.

Beeftink, W.G., 1977b. The coastal saltmarshes of western and northern Europe: an ecological and phytosociological approach. In: *Wet Coastal Ecosystems*, V.J. Chapman ed., 109–155. Elsevier, Amsterdam.

Bell, M., Caseldine, A. & Neumann, H., 2000. *Prehistoric Intertidal Archaeology in the Welsh Severn Estuary*. Council for British Archaeology.

Berg, G., Esselink, P., Groeneweg, M. & Kiehl, K., 1997. Micropatterns in *Festuca rubra* dominated salt-marsh vegetation induced by sheep grazing. *Plant Ecology*, **132/1**, 1–14.
Bernhardt, K.G. & Koch. M., 2003. Restoration of a saltmarsh system: temporal change of plant species diversity and composition. *Basic and Applied Ecology*, **4/5**, 441–451.
Bertness, M.D., 1991. Zonation of *Spartina patens* and *Spartina alterniflora* in New England Saltmarsh. *Ecology*, **72/1**, 138–148.
Best, S., 2004. *A Whim Set in Concrete: The Campaign to Stop the Cardiff Bay Barrage*. Seren Books, Bridgend.
Bird, E.C.F., 1984. *Coasts – an Introduction to Coastal Geomorphology*. 3rd Edition, Basil Blackwell, Oxford.
Bird, E.C.F., 1985. *Coastline Changes: a Global Review*. John Wiley & Sons, Chichester.
Bixing, S. & Phillips, S.M., 2006. 147 *Spartina*, Schreber, Gen. Pl. 1789. *Flora of China*, **22**, 493–494.
Blew, J. & Südbeck, P., eds., 2005. *Migratory Waterbirds in the Wadden Sea 1980 – 2000*. Wadden Sea Ecosystem, **20**, Common Wadden Sea Secretariat, Trilateral Monitoring and Assessment Group, Joint Monitoring Group of Migratory Birds in the Wadden Sea, Wilhelmshaven, Germany. Source: http://www.waddensea-secretariat.org/TMAP/wse20/wse20.html.
Blondel, J. & Aronson, J., 1999. *Biology and Wildlife of the Mediterranean Region*. Oxford University Press, Oxford.
Boesch, D.F. & Turner, R.E., 1984. Dependence of fishery species on saltmarshes: the role of food and refuge. *Estuaries*, **7/4**, Part A: Faunal Relationships in Seagrass and Marsh Ecosystems, 460–468.
Boorman, L.A. & Ashton, C., 1997. The productivity of saltmarsh vegetation at Tollesbury, Essex and Stiffkey, Norfolk, England. *Mangroves and Saltmarshes*, **1**, 113–126.
Boorman, L.A. & Ranwell, D.A., 1977. *Ecology of Maplin Sands and the Coastal Zones of Suffolk, Essex and North Kent*. Institute of Terrestrial Ecology, Natural Environment Research Council, Cambridge.
Boorman, L.A., 1999. Saltmarshes – present functioning and future change. *Mangroves and Saltmarshes*, **3/4**, 227–241.
Boorman, L.A., 2003. *Saltmarsh Review. An overview of coastal saltmarshes, their dynamic and sensitivity characteristics for conservation and management*. JNCC Report, No. **334**, JNCC, Peterborough.
Bortolus, A., 2006. The austral cordgrass *Spartina densiflora* Brong : its taxonomy, biogeography and natural history. *Journal of Biogeography*, **33/1**, 158–168.
Bos, D., Bakker, J.P., de Vries, Y. & van Lieshout, S., 2002. Long term vegetation changes in experimentally grazed and ungrazed barrier marshes in the Wadden Sea. *Applied Vegetation Science*, **5/1**, 45–54.
Boston, K.G., 1983. The development of salt pans on tidal marshes, with particular reference to South-Eastern Australia. *Journal of Biogeography*, **10/1**, 1–10.
Bouma, T.J., De Vries, M.B., Low, E., Peralta, G., Tánczos, I.C., van de Koppel, J. & Herman, P.M.J., 2005. Trade offs related to ecosystem engineering: a case study on stiffness of emerging macrophytes. *Ecology*, **86/8**, 2187–2199.
Brampton, A., 1992. Engineering significance of British saltmarshes. In: *Saltmarshes: Morphodynamics, Conservation and Engineering Significance*, J.R.L. Allen & K. Pye, eds., 115–122. Cambridge University Press, Cambridge.
Brew, D.S. & Williams, A., 2002. Shoreline Movement and Shoreline Management in The Wash, Eastern England. In: *The Changing Coast, Littoral 2002*. EUROCOAST / EUCC, Porto - Portugal, 313– 320.
Brindley, E., Norris, K., Cook, T., Babbs, S., Forster Brown, C., Massey, P., Thompson, R. & Yaxley, R., 1998. The abundance and conservation status of redshank *Tringa totanus* nesting on saltmarshes in Great Britain. *Biological Conservation*, **86/3**, 289–297.
Brochier, F. & Ramieri, E., 2001. *Climate Change Impacts on the Mediterranean Coastal Zones*. Climate Change, Modelling and Policy, Note di Lavoro Series, Fondazione Eni

Enrico Mattei. Source: http://www.feem.it/NR/rdonlyres/2F94BCDC-9344-4FC6-B0D9-7DCD642B23FC/549/2701.pdf.

Brooke, J., Landin, M., Meakins, N. & Adnitt, C., 1999. The Restoration of Vegetation on Saltmarshes. *Environment Agency R&D Technical Report*, **208**, Environment Agency.

Broome, S.W. 1990. Creation and restoration of tidal wetlands of the southeastern United States. In: *Wetland Creation and Restoration: The Status of the Science*, J. A. Kusler & M. E. Kentula, eds., 37–72. Island Press, Washington, D.C.

Broome, S.W., Seneca, E.D. & Woodhouse, W.W. Jr., 1986. Long-term growth and development of transplants of the salt-marsh grass *Spartina alterniflora*. *Estuaries*, **9/1**, 63–74.

Burd, F., 1989. *The Saltmarsh Survey of Great Britain. An Inventory of British Saltmarshes*. Research & survey in nature conservation, **17**, Nature Conservancy Council, Peterborough.

Burd, F., 1992. *Erosion and Vegetation Change on the Saltmarshes of Essex and North Kent between 1973 and 1988*. Research & survey in nature conservation, **42**, Nature Conservancy Council, Peterborough.

Burd, F., 1994. *Sites of Historical Sea Defence Failure - Phase II Study*. A Report to English Nature, Peterborough, Hull: Institute of Estuarine and Coastal Studies.

Cabanes, C., Cazenave, A. & Le Provost, C., 2001. Sea level rise during the past 40 Years determined from satellite and in situ observations. *Science*, **294/5543**, 840–842.

Cadbury, C.J., Green, R.E. & Allport, G., 1987. Redshank and other breeding waders of British saltmarshes. RSPB Conservation Review, 1, 37–39.

Cadwalladr, D.A. & Morley, J.V., 1971. Further experiments on the management of saltings pasture for Wigeon (*Anas penelope* L.) conservation at Bridgwater Bay National Nature Reserve, Somerset. *Journal of Applied Ecology*, **10**, 161–166.

Cadwalladr, D.A., Owen, M., Morley, J.V. & Cook, R.S., 1972. Wigeon (*Anas penelope*) conservation and salting pasture management at Bridgwater Bay National Nature Reserve, Somerset. *Journal of Applied Ecology*, **9**, 117–126.

Callaway, J.C. & Josselyn, M.N., 1992. The introduction and spread of Smooth Cordgrass (*Spartina alterniflora*) in South San Francisco Bay. *Estuaries*, **15/2**, 218–226.

Carey, A.E. & Oliver, F.W., 1918. *Tidal Lands, a Study in Shore Problems*. Blackie and Son, London.

Cargill, S.M. & Jefferies, R.L., 1984. The effects of grazing by lesser snow geese on the vegetation of a subarctic saltmarsh. *Journal of Applied Ecology*, **21**, 669–686.

Carter, R.W.G. & Woodroffe, C.D., 1994. *Coastal Evolution – Late Quaternary Shoreline Morphodynamics*. Cambridge University Press, Cambridge.

Carter, R.W.G., 1988. *Coastal Environments. An Introduction to the Physical, Ecological and Cultural Systems of Coastlines*. Academic Press, London.

Castillo, J.M., Fernandez-Baco, L., Castellanos, E.M., Luque, Figueroa, M.E. & Davy, A.J., 2000. Lower limits of *Spartina densiflora* and *S. maritima* in a Mediterranean saltmarsh determined by different ecophysiological tolerances. *Journal of Ecology*, **88**, 801–812.

Castillo, J.M., Rubio-Casal, A.E., Luque, C.J., Nieva, F.J. & Figueroa, M.E., 2002. Wetland loss by erosion in Odiel Marshes (SW Spain), *Journal of Coastal Research*, **36**, 134–138.

Catry, T., Ramos, J.A., Catry, I., Allen-Revez, M. & Grade, N., 2004. Are salinas a suitable alternative breeding habitat for Little Terns *Sterna albifrons*? *Ibis*, **146/2**, 247–254.

Cazenave, A., Cabanes, C., Dominh, K. & Mangiarotti, S. 2001. Recent sea level change in the Mediterranean Sea revealed by Topex/Poseidon satellite altimetry. *Geophysical Research Letters*, **28/8**, 1607–1610.

Chapman, V.J., 1938. Studies in saltmarsh ecology I–III. Physiography and environmental factors; the tides; the water table, soil aeration and drainage. *Journal of Ecology*, **26**, 144–179.

Chapman, V.J., 1939. Studies in saltmarsh ecology V, the vegetation. *Journal of Ecology*, **27**, 181–201.

Chapman, V.J., 1941. Studies in saltmarsh ecology VIII. *Journal of Ecology*, **29**, 69–82.

Chapman, V.J., 1974. *Saltmarshes and Salt Deserts of the World*. Leonard Hill, London.

Chapman. V.J., 1960. The plant ecology of Scolt Head Island. In: J.A. Steers, ed., *Scolt Head Island*, Heffer, Cambridge, 85–163.

Charman, K., 1990. The current status of *Spartina anglica* in Britain. In: *Spartina anglica - a Research Review*, A.J. Gray & P.E.M. Benham, eds., 11–14. Institute of Terrestrial Ecology, HMSO, London.

Chen, Z., Li, B., Zhong, Y. & Chen, J., 2004. Local competitive effects of introduced *Spartina alterniflora* on *Scirpus mariqueter* at Dongtan of Chongming Island, the Yangtze River estuary and their potential ecological consequences. *Hydrobiologia*, **528/1–3**, 99–106.

Chesapeake Bay Nutria Working Group 2003. *Nutria (Myocastor coypus) in the Chesapeake Bay: Draft Bay-Wide Management Plan*. Source: http://www.chesapeakebay.net/pubs/calendar/NISW_12-10-03_Report_5_5129.pdf.

Chmura, G.L. Helmer, L.L. Beecher, C.B. & Sunderland, E.M., 2001. Historical rates of saltmarsh accretion on the outer Bay of Fundy. *Canadian Journal of Earth Science*, **38**, 1081–1092.

Chung, C-H., 1990. Twenty-five years of introduced *Spartina anglica* in China. In: *Spartina anglica - a Research Review*, A.J. Gray & P.E.M. Benham, eds., 72–76. Institute of Terrestrial Ecology, HMSO, London.

Chung, C-H., 1993. Thirty years of ecological engineering with *Spartina* plantations in China. *Ecological Engineering*, **2/3**, 261–289.

Clark, N.A., 2006. Tidal barrages and birds. *Ibis*, **148**, 152–157.

Coastal Geomorphological Partnership, 2001. *Coastal Data Analysis: The Wash. Study 2: Wave Attenuation over Inter-Tidal Surfaces*. Environment Agency (Anglian Region).

Conway, C.J., Sulzman, C. and Raulston, B.A., 2004. Factors affecting detection probability of California Black Rails. *Journal of Wildlife Management*, **68**, 360–370.

Cooper, N.J., Cooper, T. & Burd, F., 2001. 25 years of saltmarsh erosion in Essex. Implications for coastal defence and nature conservation. *Journal of Coastal Conservation*, **7/1**, 31–40.

Cooper, N.J., Skrzypczak, T. & Burd, F., 2000. *Erosion of the Saltmarshes of Essex between 1988 and 1998. Volume 1: Implications for Flood Defence and Nature Conservation; Volume 2: A Recommended Monitoring Framework for the future; Volume 3: Maps of Changes in the Essex Saltmarshes, 1988–1998*. Report to the Environment Agency (Anglian Region).

Corkhill, P., 1984. *Spartina* at Lindisfarne NNR and details of recent attempts to control its spread. In: *Spartina anglica in Great Britain*, J.P. Doody, ed., 60–63. Focus on nature conservation, **5**, Nature Conservancy Council, Attingham Park.

Costa, M.J., Catarino, F. & Bettencourt, A., 1988. The role of saltmarshes in the Mira estuary (Portugal). *Wetlands Ecology and Management*, **9/2**, 121–134.

Crooks, S., Schutten, J., Sheem, G.D., Pye, K. & Davy, A.J., 2002. Drainage and elevation as factors in the restoration of saltmarsh in Britain. *Restoration Ecology*, **10**, 591–602.

Curtis, T.G.F. & Sheehy Skeffington, M.J., 1998. The saltmarshes of Ireland an inventory and account of their geographical variation, biology and environment. *Proceedings of the Royal Irish Academy*, **98B/2**, 87–104.

d'Angremond, K., 2003. From Disaster to Delta Project: The Storm Flood of 1953. *Terra et Aqua*, **90**, 3–10.

Daehler, C.C. & Strong, D.R., 1997a. Hybridization between introduced smooth cordgrass (*Spartina alterniflora*; Poaceae) and native California cordgrass (*S. Foliosa*) in San Francisco Bay, California, USA. *American Journal of Botany*, **84+A902/5**, 607–611.

Daehler, C.C. & Strong, D.R., 1997b. Unusual potential for biological control of alien cordgrasses in Willapa bay. *Proceedings of the Second International Spartina Conference*. Olympia WA. 20–21 March, 1997.

Dagley, J.R., 1995. *Northey Island: Managed Retreat Scheme*. English Nature Research Report, **128**, English Nature, Peterborough.

Dahl, T.E. & Allord, G.J. 1997. *Technical Aspects of Wetlands: History of Wetlands in the Conterminous United States*. National Water Summary on Wetland Resources, United States

Geological Survey, Water Supply Paper, 2425. Source: http://water.usgs.gov/nwsum/WSP2425/history.html.

Dai, T. & Wiegert, R.G., 1996. Estimation of the primary productivity of *Spartina alterniflora* using a canopy model. *Ecography* **19/4**, 410–423.

Daiber, F.C., 1986. *Conservation of Tidal Marshes*. Van Nostrand Reinhold, New York.

Daily, G.C., Alexander, S., Ehrlich, P.R., Goulder, L., Lubchenco, J., Matson, P.A., Mooney, H.A., Postel, S., Schneider, S.H., Tilman, D. & Woodwell, G.M., 1997. Ecosystem services: benefits supplied to human societies by natural ecosystems. *Issues in Ecology*, **1/2**, 1–18.

Dalby, R., 1957. Problems of land reclamation, 5, Saltmarsh in the Wash. *Agricultural Review*, London, **2**, 31–37.

Davidson, N.C., Laffoley, D.d'A., Doody, J.P., Way, L.S., Gordon, J., Key, R., Drake, C.M., Pienkowski, M.W., Mitchell, R. & Duff, K.L., 1991. *Nature Conservation and Estuaries in Great Britain*. Nature Conservancy Council, Peterborough.

Davis. P. & Moss, D., 1984. *Spartina* and waders the Dyfi estuary. In: *Spartina anglica in Great Britain*, J.P Doody, ed., 37–40. Focus on nature conservation, **5**, Nature Conservancy Council, Attingham Park.

Davison, D.M. & Hughes, D.J., 1998. *Zostera Biotopes (volume I). An Overview of Dynamics and Sensitivity Characteristics for Conservation Management of Marine SACs*. Scottish Association for Marine Science (UK Marine SACs Project).

Day, J.W. Jr., Pont, D., Hensel, P.F. & Ibañez, C., 1995. Impacts of Sea-Level Rise on Deltas in the Gulf of Mexico and the Mediterranean: The Importance of Pulsing Events to Sustainability. *Estuaries*, **18/4**, Papers from William E. Odum Memorial Symposium (Dec.), 636–647.

Day, J.W., Jr., Scarton, F., Rismondo, A. & Are, D., 1998. Rapid Deterioration of a Saltmarsh in Venice Lagoon, Italy. *Journal of Coastal Research*, **14/2**, 583–590.

Deboeuf, C. & Herrier, J.-L., 2002. The restoration of mudflats, saltmarshes and dunes on the eastern bank of the Yzer-rivermouth, Nieuwpoort. In: *Littoral 2002, the Changing Coast*, ed., Eurocoast – Portugal, Eurocoast/EUCC, Porto – Portugal, Volume III, Posters, 201–202.

Denny, M.J.H. & Anderson, G.Q.A., 1999. *Effects of rotoburying Spartina anglica at Lindisfarne NNR. Results for plant species, two years after rotoburying*. Report to English Nature.

Department of Environment, Food and Rural Affairs, 2001. *Shoreline Management Plans: A Guide for Coastal Defence Authorities*. Department of Environment, Food and Rural Affairs, UK. Continuously updated. Source: http://www.defra.gov.uk/environ/fcd/policy/SMP.htm#smp1

Department of the Interior, US Fish and Wildlife Service, Willapa National Wildlife Refuge, 1997. *Control of Smooth Cordgrass (Spartina alterniflora) on Willapa National Wildlife Refuge: Environmental Assessment*. Source: http://www.spartina.org/project_documents/washington/ea_full.pdf.

Department of Trade, 1974. *Maplin: review of airport project*. HMSO. Source: http://www.bopcris.ac.uk/bopall/ref16472.html.

Dierschke, J. & Bairlein, F., 2004. Habitat selection of wintering passerines in saltmarshes of the German Wadden Sea. *Journal of Ornithology*, **145/1**, 48–58.

Dijkema, K.S. & Wolf, W.J., 1983. *Flora and Vegetation of the Wadden Sea Islands and Coastal Areas*. Report **9** of the Wadden Sea Working Group, Stitching Veth tot Steun aan Waddenonderzoek, Leiden, 305–308.

Dijkema, K.S. ed., 1984. *Saltmarshes in Europe*. Nature and Environment series, **30**, Council of Europe, Strasbourg.

Dijkema, K.S., 1990. Salt and brackish marshes around the Baltic Sea and adjacent parts of the North Sea: their vegetation and management. *Biological Conservation*, **51**, 191–209.

Dijkema, K.S., 1987. Changes of salt-marsh area in the Netherlands Wadden Sea after 1600. In: *Vegetation between land and sea*, A.H.L., Huiskes, C.W.P.M. Blom, & J. Rozema, eds., 42–49. Junk, Dordrecht.

Dijkema, K.S., 1997. Impact prognosis for saltmarshes from subsidence by gas extraction in the Wadden Sea. *Journal of Coastal Research*, **13/4**, 1294–1304.

Dijkema, K.S., de Jong, D.J., Vreeken-Buijs, M.J. & van Duin, W.E., 2005. *Saltmarshes in the Water Framework Directive. Development of Potential Reference Conditions and of Good Ecological Statuses*. Alterra-Texel, Rijkwaterstaat RIKZ, Report 2005.020.
Doody, J.P. & Barnet, B., 1987. *The Wash and its Environment*. Research and survey in nature conservation, **7**, Peterborough, Nature Conservancy Council.
Doody, J.P. ed., 1984. *Spartina anglica in Great Britain*. Report of a meeting held at Liverpool University, 10 November, 1982. Nature Conservancy Council, Attingham Park.
Doody, J.P., 1987. The impact of 'reclamation' on the natural environment of the Wash. In: *The Wash and its environment*, P. Doody & B. Barnett, eds., 165–172. Research &survey in nature conservation, **7**, Nature Conservancy Council, Peterborough.
Doody, J.P., 1990. Spartina- friend or foe? A conservation viewpoint. In: *Spartina anglica a Research Review*, eds., A.J. Gray & P.E.M. Benham, 77–79. HMSO, London.
Doody, J.P., 1992. Sea defence and nature conservation: threat or opportunity. *Journal of Aquatic Conservation*, **2**, 275–283.
Doody, J.P., 2001. *Coastal Conservation and Management: an Ecological Perspective*. Kluwer, Academic Publishers, Boston, USA, 306 pp. *Conservation Biology Series*, 13
Doody, J.P., 2004. 'Coastal squeeze' – an historical perspective. *Journal of coastal Conservation*, **10/1–2**, 129–138.
Doody, J.P., Ferreira, M., Lombardo, S., Lucius, I., Misdorp, R., Niesing, H., Salman, A. & Smallegange, M., 2004. Living with Coastal Erosion in Europe – Sediment and Space for Sustainability. Results from the EUROSION Study. European Commission, Office for Official Publications of the European Communities. Source: http://www.eurosion.org/project/eurosion_en.pdf.
Douglas, B.C., 1991. Global sea level rise. *Journal of Geophysical Research*, **96C/4**, 6981–6992.
Douglas, B.C., 1997. Global sea rise: a redetermination. *Surveys in Geophysics* **18/2–3**, 279–292.
Drent, R.H. & Van Der Wal, R., 1999. Cyclic grazing invertebrates and the manipulation of the food resource. In: *Herbivores between plants and predators*, H. Olff, V.K. Brown, & R.H. Drent, eds., 271–299. Blackwell Science, Oxford.
Dugdale, R., Plater, A. & Albanakis. K., 1987. The fluvial and marine contribution to the sediment budget of the Wash. In: *The Wash and its Environment*, J.P. Doody, & B. Barnet, eds. 37–47. Research and survey in nature conservation, **7**, Nature Conservancy Council. Peterborough.
Duke, J.A., 1983. *Handbook of Energy Crops*. Unpublished. Information source: New Crop Resource Online Program, http://www.hort.purdue.edu/newcrop/duke_energy/Phragmites_australis.html
Eddleman, W.R., Flores, R.E. & Legare, M.L., 1994. Black Rail (*Laterallus jamaicensis*). *In*: *The Birds of North America, 123*, A. Poole & F. Gill, eds. The Academy of Natural Sciences, Philadelphia.
Edmonson, S.E., Traynor, H. & McKinnell, S., 2001. The development of a green beach on the Sefton Coast, Merseyside, UK. In: *Coastal Dune Management, Shared Experience of European Conservation Practice*, J.A. Houston, S.E. Edmondson, & P.J. Rooney, eds., 48–58. Liverpool University Press.
Edwards, K.R., & Mills, K.P. 2005. Aboveground and Belowground Productivity of *Spartina alterniflora* (Smooth Cordgrass) in Natural and Created Louisiana Saltmarshes. *Estuaries*, **28/2**, 252–265.
Edwards, K.R. & Proffitt, C.E., 2003. Comparison of wetland structural characteristics between created and natural saltmarshes in southwest Louisiana, USA. *Wetlands*, **23/2**, 344–356.
Eertman, R.H.M., Kornman, B.A., Stikvoort, E. & Verbeek, H., 2002. Restoration of the Sieperda Tidal Marsh in the Scheldt Estuary, The Netherlands. *Restoration Ecology*, **10/3**, 438–449.
Esselink, P., Fresco, Latzi, F.M. & Dijkema, K.S., 2001 Vegetation change in a man-made saltmarsh affected by a reduction in both grazing and drainage. *Applied Vegetation Science*, **5/1**, 17–32.
Esselink, P., Zijlstra, W., Dijkema, K.S. & van Diggelen, R., 2000. The effects of decreased management on plant-species distribution patterns in a saltmarsh nature reserve in the Wadden Sea. *Biological Conservation*, **93/1**, 61–76.

Estuary Habitat Restoration Partnership Act, 1999. Senate Report 106–189 – Calendar No. 323, 106th CONGRESS. Source: http://www.congress.gov/cgi-bin/cpquery/R?cp106:FLD010:@1(sr189)

European Commission, 1999. *Towards a European Integrated Coastal Zone Management (ICZM) Strategy: General Principles and Policy Options. A Reflection Paper.* Directorates-General – Environment, Nuclear Safety & Civil Protection; Fisheries and Regional Policy & Cohesion. Source: http://ec.europa.eu/environment/iczm/pdf/vol1.pdf.

European Commission, 2003. *Interpretation Manual of European Habitats. Natura 2000.* European Commission, DG Environment, Nature and Biodiversity, Brussels. Source: http://ec.europa.eu/environment/nature/nature_conservation/eu_enlargement/2004/pdf/habitats_im_en.pdf.

Evans, G. & Collins, M., 1987. Sediment supply and deposition in the Wash. In: *The Wash and its Environment*, J.P. Doody & B. Barnett, eds., 58–63. Nature Conservancy Council, Peterborough.

Evans, P.R., 1984. Use of the herbicide 'Dalapon' for control of *Spartina* encroaching on intertidal mudflats: beneficial effects on shorebirds. *Colonial Waterbirds*, **9**, 171–175.

Fasham, M. & Trumper, K., 2001. *Review of Non-native Species Legislation and Guidance.* Report to the Department for Environment, Food & Rural Affairs. P328 DEFRA NNS review V5.doc. Source: http://www.defra.gov.uk/wildlife-countryside/resprog/findings/non-native/index.htm.

Fasola, M. & Ruiz, X., 1996. The value of rice fields as substitutes for natural wetlands for waterbirds in the Mediterranean Region. *Colonial Waterbirds*, **19**, Special issue, 122–128.

Fasola, M., Canova, L. & Saino, N., 1996. Rice fields support a large portion of herons breeding in the Mediterranean Region. *Colonial Waterbirds*, **19**, Special issue, 129–134.

Feist, B.E. & Simenstad, C.A., 2000. Expansion rates and recruitment frequency of exotic Smooth Cordgrass, *Spartina alterniflora* (Loisel), colonizing unvegetated littoral flats in Willapa Bay, Washington. *Estuaries*, **23/2**, 267–274.

Field, D.W., Reyer, A.J., Genovese, P.V. & Shearer, B.D., 1991. *Coastal Wetlands of the United States.* National Oceanographic and Atmospheric Administration and US Fish and Wildlife Service, Washington, DC.

Flessa, K.W. Constantine, K.J. & Cushman, M.K., 1977. Sedimentation Rates in a Coastal Marsh Determined from Historical Records. *Chesapeake Science*, **18/2**, 172–176.

Fox, A.D., Kahlert, J. & Ettrup, H., 1998. Diet and habitat use of moulting Greylag Geese *Anser anser* on the Danish island of Saltholm. *Ibis*, **140/4**, 676–683.

Fraser, S., 1999. *RAMSAR Sites Overview. A Synopsis of the World's Wetlands of International Importance.* Wetlands International.

French, J.R. & Burningham, H., 2003. Tidal marsh sedimentation versus sea level rise: a southeast England perspective. *Proceedings Coastal Sediments '03*, Sheraton Sand Key, Clearwater, Florida, 1–14.

French, P.W., 1997. *Coastal and Estuarine Management.* Routledge Environmental Management Series, London.

Frey, R.W. & Bason, P.B., 1978. North American saltmarshes. In: *Coastal Sedimentary Environments*, R.A.A. Davis, ed., 104–164. Springer, Berlin.

Frid, C.L.J., Chandrasekara, W.U. & Davey, P., 1999. The restoration of mud flats invaded by common cord-grass (*Spartina anglica*, CE Hubbard) using mechanical disturbance and its effects on the macrobenthic fauna. *Aquatic Conservation: Marine and Freshwater Ecosystems*, **9/1**, 47–61.

Gallagher, J.L. & Plumley, F.G., 1979. Underground Biomass Profiles and Productivity in Atlantic Coastal Marshes. *American Journal of Botany*, **66/2**, 156–161.

Gallois, R.W., 1994. Geology of the Country around King's Lynn and the Wash. *Memoirs of the Geological Survey*, H.M.S.O., London.

Ganter, B., Prokosch, P. & Ebbinge, B.S., 1998. Effect of saltmarsh loss on the dispersal and fitness parameters of Dark-bellied Brent Geese. *Aquatic Conservation: Marine and Freshwater Ecosystems*, **7/2**, 141–151.

Garbisch, E., 1994. The results of muskrat feeding on cattails in a tidal freshwater wetland. *Wetland Journal*, **6/1**, 14–15.

Garbutt, R.A., Reading, C.J., Wolters, M., Gray, A.J., & Rothery, P., 2006. Monitoring the development of intertidal habitats on former agricultural land after the managed realignment of coastal defences at Tollesbury, Essex, UK. *Marine Pollution Bulletin*, **53/1–4**, 155–164.

Garniel, A. & Mierwald, U., 1996. Changes in the morphology and vegetation along the human altered shoreline of the Lower Elbe. In: K.F. Nordstrom & C.T. Roman, *Estuarine Shores – Evolution, Environments and Human* Activities, 375–396. John Wiley & Sons, New York.

Gehrels, W.R., Belknap, D.F., Black, S. & Newnham, R.M., 2002. Rapid sea-level rise in the Gulf of Maine, USA, since AD 1800. *The Holocene*, **12/4**, 383–389.

Goeldner, L., 1999. The German Wadden Sea coast: reclamation and environmental protection. *Journal of Coastal Conservation*, 5, 23–30.

Goodman, P.J., 1960. Investigations into 'die-back' in *Spartina townsendii agg*. II. The morphological structure and composition of the Lymington sward. *Journal of Ecology*, **48**, 711–725.

Goodman, P.J., 1969. *Spartina* Schreb. Biological Flora of the British Isles. *Journal of Ecology*, **57/1**, 285–287.

Goodman, P.J., Braybrooks, E.M. & Lambert, J.M., 1959. Investigations into 'die-back'. In: *Spartina townsendii agg*. 1: The present status of *Spartina townsendii* in Britain. *Journal of Ecology*, **47**, 651–677.

Goodman, P.J., Braybrooks, E.M., Marchant, C.J. & Lambert, J.M., 1969. *Spartina X Townsendii*, H. & J. Groves Sensu Lato. *The Journal of Ecology*, **57/1**, 198–313.

Gosling, L.M. & Baker, S.J., 1989. The eradication of muskrats and coypus from Britain. *Biological Journal of the Linnean Society*, **38/1**, 39–51.

Goss-Custard & Moser, M., 1988. Rates of change in the numbers of Dunlin (*Calidris alpina*) wintering in British estuaries in relation to the spread of *Spartina anglica*. *Journal of Applied Ecology*, **25/1**, 95–109.

Gray, A.J. & Benham, P.E.M., 1990. *Spartina anglica – a research review*. Institute of Terrestrial Ecology, HMSO, London.

Gray, A.J. & Pearson. J.M., eds., 1984. *Spartina* marshes in Poole Harbour, Dorset with particular reference to Holes Bay. In*: Spartina anglica in Great Britain*, J.P Doody, ed., 11–14. Focus on nature conservation, **5**, Nature Conservancy Council, Attingham Park.

Gray, A.J. & Raybould, A.F. 1997. The history and evolution of *Spartina anglica* in the British Isles. *Proceedings of the Second International Spartina Conference*, 1997, 12–16.

Gray, A.J., 1972. The ecology of Morecambe Bay. V. The Saltmarshes of Morecambe Bay. *Journal of Applied Ecology*, **9**, 207–220.

Gray, A.J., 1977. Reclaimed land. In: *The Coastline*, ed., R.S.K. Barnes, John Wiley & Sons, Chichester, 253–270.

Gray, A.J., 1992. Saltmarsh plant ecology. In: *Saltmarshes: morphodynamics, conservation and engineering* significance, J.R.L., Allen & K. Pye, eds., 63–79. Cambridge University Press, Cambridge.

Gray, A.J., Benham, P.E.M. & Raybould, A.F., 1990. *Spartina anglica* – the evolutionary and ecological background. In: *Spartina anglica – a Research Review*, A.J. Gray & P.E.M. Benham, eds., 5–10. Institute of Terrestrial Ecology, HMSO, London.

Gray, A.J., Marshall, D.F. & Raybould, A.F., 1991. A century of evolution in *Spartina anglica*. *Advances in Ecological Research*, **21**, 1–61.

Gray, A.J., Raybould, A.F. & Brown, S.L., 1997. The environmental impact of *Spartina anglica*: past, present and predicted. *Proceedings of the Second International Spartina Conference*. Olympia WA. March 20–21, 97.

Greenberg, R. & Droege, S., 1990. Adaptations to tidal marshes in breeding populations of the swamp sparrow. *Condor*, **92**, 393–404.

Greenberg, R., Maldonado, J., Droege, S. & McDonald, M.V., 2006. Tidal marshes: a global perspective on the evolution and conservation of their terrestrial vertebrates. *BioScience*, **56/8**, 675–685.

Grieve, H., 1959. *The Great Tide*. County Council of Essex.

Grove, A.T. & Rackham, O., 2001. *The Nature of the Mediterranean Europe: an Ecological History*. Yale University Press, New Haven and London.

Hacker, S.D., Heimer, D., Hellquist, C.E., Reeder, T.G., Reeves, B., Riordan, T.J. & Dethier, M.N., 2001. A marine plant (*Spartina anglica*) invades widely varying habitats: potential mechanisms of invasion and control. *Biological Invasions*, **3/2**, 211–217.

Hammond, M.E.R. & Cooper, A., 2002. *Spartina anglica* eradication and inter-tidal recovery in Northern Ireland estuaries. In: *Turning the Tide: the Eradication of Invasive Species*, C.R. Veitch & M.N. Clout, eds., 124–131. IUCN, SSC, Invasive Species Specialist Group, IUCN Gland, Switzerland and Cambridge, UK.

Handa, I.T. & Jefferies, R.L., 2000. Assisted revegetation trials in degraded salt-marshes. *Journal Applied Ecology*, **37/6**, 944–958.

Hannaford, J., Pinn, E.H. & Diaz, A., 2006. The impact of sika deer grazing on the vegetation and infauna of Arne saltmarsh. *Marine Pollution Bulletin*, **53/1–4**, 56–62.

Harmsworth, G.C. & Long, S.P., 1986. An assessment of saltmarsh erosion in Essex, England, with reference to the Dengie Peninsula. *Biological Conservation*. **35/4**, 377–387.

Hartig, E.K. & Gornitz, V., 2001. The Vanishing Marshes of Jamaica Bay: Sea Level Rise or Environmental Degradation? *Goddard Institute of Space Studies, Science Briefs*. Source: http://www.giss.nasa.gov/research/briefs/hartig_01/.

Hartig, E.K., Gornitz, V., Kolker, A., Mushacke, F. & Fallon, D., 2002. Anthropogenic and climate change impacts on saltmarshes of Jamaica Bay, New York City. *Wetlands*, **22/1**, 71–89.

Haslett, S.K., Cundy, A.B., Davies, C.F.C., Powell, E.S. & Croudacec, I.W., 2003. Saltmarsh sedimentation over the past c. 120 years along the West Cotentin coast of Normandy (France): relationship to sea-level rise and sediment supply. *Journal of Coastal Research*, **19/3**, 609–620.

Haynes. F.N., 1984. *Spartina* in Langstone Harbour, Hampshire. In: *Spartina anglica in Great Britain*, J.P. Doody, ed., 5–10. Focus on nature conservation, No. **5**, Nature Conservancy Council, Attingham Park.

Hearn, R.D., 2004. Greater White-fronted Goose *Anser albifrons albifrons* (Baltic-North Sea population) in Britain 1960/1961 – 1999/2000. *Waterbird Review Series*, the Wildfowl & Wetlands Trust/Joint Nature Conservation Committee, Slimbridge.

Hedge, P. & Kriwoken, L., 1997. Managing *Spartina* in Victoria and Tasmania, Australia. In: *Second International Spartina Conference*: *Proceedings*, Olympia WA, 93–96.

Hedge, P., 2002. Strategy for the Management of Rice Grass (*Spartina anglica*) in Tasmania, Australia, (2nd edition). Rice Grass Advisory Group (RGAG), Rice Grass Management Programme, Hobart, Tasmania. Source: http://www.dpiw.tas.gov.au/inter.nsf/WebPages/ALIR-4Z587L?open.

Hedge, P., Kriwoken, L. & Ritar, A., 1997. The distribution of *Spartina* in Victoria and Tasmania, Australia. *Second International Spartina Conference*: *Proceedings*, Olympia WA, 9–11.

Hedge, P., Kriwoken, L.K. & Patten, K., 2003. A review of *Spartina* management in Washington State, US. *Journal of Aquatic Plant Management*, **41**, 82–90.

Herrier, J.-L., van Nieuwenhuyse, H., Deboeuf, C., Deruyter, S. & Leten, M., 2005. Sledgehammers, cranes and bulldozers: restoring dunes and marshes by removing buildings and soil. In: *Proceedings 'Dunes and Estuaries 2005' – International Conference on Nature Restoration Practices in European Coastal Habitats*, J.-L. Herrier, J. Mees, A. Salman, J. Seys, H. Van Nieuwenhuyse & I. Dobbelaere, eds., 79–94. Koksijde, Belgium, 19–23 September, 2005. VLIZ Special Publication **19**.

Hill, M. I., 1988. *Saltmarsh Vegetation of the Wash – an Assessment of Change from 1971 to 1987*. Research & survey in nature conservation, **13**, Nature Conservancy Council, Peterborough.

Hill, M.I. & Randerson, P.R., 1987. Saltmarsh vegetation communities of the Wash and their recent development. In: *The Wash and Its Environment*, P. Doody & B. Barnet, eds., 111–122. Research & survey in nature conservation, **7**, Nature Conservancy Council, Peterborough.

Hofstede, J., 2004. A new coastal defence master plan for Schleswig-Holstein. *Coastline Reports*, **1**, 109–117. Source: http://www.eucc-d.de/plugins/coastline_reports/pdf/cr1/AMK2004_Artikel_Hofstede.pdf

Holder, C. L. & Burd, F., 1990. *Overview of Saltmarsh Restoration Sites in Essex*. Interim report. Nature Conservancy Council, Contract surveys, **83**, Peterborough.

Hötker, H. 1997. Response of migratory coastal bird populations to the land claim in the Nordstrand Bay, Germany. In: *Effect of habitat loss and change on waterbirds*, J.D. Goss-Custard, R. Rufino & A. Luis, eds., 52–62. ITE Symposium, **30**, London: The Stationery Office.

Hough, A., Spencer, C., Lowther, S. & Muddiman, S., 1999. *Definition of the Extent and Vertical Range of Saltmarsh*. Environment Agency, Technical Report W **153**. Research Contractor, SGS Environment.

Houghton, J.T.Y., Griggs, D.D.J., Noguer, M., van der Linden, P.J., Dai, X., Maskell, K. & Johnson, C.A. 2001. *Climate Change 2001: The Scientific Basis. Contribution of Working Group I to the Third Assessment Report of the Intergovernmental Panel on Climate Change*. Cambridge University Press, Cambridge, United Kingdom and New York, NY, USA. Source: http://www.grida.no/climate/ipcc_tar/wg1/index.htm.

Houwing, E.J., van Duin, W.E., Smit-van der Waaij, Y., Dijkema, K.S. & Terwindt, J.H.J., 1999. Biological and abiotic factors influencing the settlement and survival of *Salicornia dolichostachya* in the intertidal pioneer zone. *Mangroves and Saltmarshes*, **3/4**, 197–206.

Howes, B.L. Weiskel, P.K., Goehringer, D.D. & Teal, M.T., 1996. Interception of freshwater and nitrogen transport from uplands to coastal waters: the role of saltmarshes. In: *Estuarine Shores – Evolution, Environments and Human Activities*, K.F. Nordstrom & C.T. Roman, eds., 287–310. John Wiley & Sons, New York.

Hubbard, J.C.E. & Partridge, T.R., 1981. Tidal immersion and the growth of *Spartina anglica* marshes in the Waihopai River Estuary, New Zealand. *New Zealand Journal of Botany*, **19**, 115–121.

Hubbard, J.C.E. & Stebbings, R.E., 1967. Distribution, dates of origin and acreage of *Spartina townsendii* (saltmarshes in Great Britain). *Transactions of the Botanical Society of the British Isles*, **7**, 1–7.

Hughes, R.G. & Paramor, O.A.L., 2004. On the loss of saltmarshes in south-east England and methods for their restoration. *Journal Applied Ecology*, **41/3**, 440–448.

Ibáñez, C., Caicio, A., Day, J.W. & Curcó, A., 1997. Morphological development, relative sea level rise and sustainable management of water and sediment in the Ebre Delta, Spain, *Journal of Coastal Conservation*, **3**, 191–202.

Ibáñez, C., 1999. Integrated management in the SPA of the Ebro Delta: implications of rice cultivation for birds. *Twenty years with the EC Birds Directive: Proceedings from a conference on the Council Directive on the Conservation of Wild Birds*. Elsinore, Denmark, 18–19 November 1999.

Ibáñez, C., Prat, N. & Canicio, A., 1996. Changes in the hydrology and sediment transport produced by large dams on the lower Ebro River and its estuary. *Regulated Rivers: Research & Management*, **12**, 51–62.

Isacch, J.P., Costa, C.S.B., Rodríguez-Gallego, L., Conde, D., Escapa, M., Gagliardini, D.A. & Iribarne, O.O., 2006. Distribution of saltmarsh plant communities associated with environmental factors along a latitudinal gradient on the south-west Atlantic coast. *Journal of Biogeography*, **33/5**, 888–900.

Jackson, D., Long, S.P. & Mason, C.F., 1986. Net primary production, decomposition and export of *Spartina anglica* on a Suffolk salt-marsh. *Journal of Ecology*, **74/3**, 647–662.

Jefferies, R.L., Rockwell, R.F. & Abraham, K.F., 2004. Agricultural food subsidies, migratory connectivity and large-scale disturbance in Arctic coastal systems: a case study. *Integrative and Comparative Biology*, **44/2**, 130–139.

Jefferies, R.L., Jano, A.P. & Abraham, K.F., 2006. A biotic agent promotes large-scale catastrophic change in the coastal marshes of Hudson Bay. *Journal of Ecology*, **94**, 234–242.

Jefferies, R.L., Rockwell, R.F. & Abraham, K.F., 2003. The embarrassment of riches: agricultural food subsidies, high goose numbers, and loss of Arctic wetlands – a continuing saga. *Environmental Review*, **11/4**, 193–232.

Jensen, A., 1985. The effect of cattle and sheep grazing on salt-marsh vegetation at Skallingen, Denmark. *Plant Ecology*, **60/1**, 37–48.

Jepsen, P.U., 1991. Crop damage and management of the Pink-footed Goose *Anser brachyrhynchus* in Denmark. *Ardea*, **79/2**, 191–194.
Jian, L., Wen-qin, D., Li-na, M., Su-fang, H. & Bo, L., 2005. An effective weed-killer for control of *Spartina anglica* C.E. Hubb. *Journal of Agro-Environment Science*, **24/2**, 410–411.
Johnston, J.B., Watzin, M.C., Barras, J.A. & Handley, L.R., 1995. Gulf of Mexico coastal wetlands: case studies of loss trends. In: *Our living resources: a report to the nation on the distribution, abundance, and health of U.S. plants, animals, and* ecosystems, E.T. LaRoe, G.S. Farris, C.E. Puckett, P.D. Doran, & M.J. Mac, eds., 269–272. US Department of the Interior, National Biological Service, Washington, DC.
Jones, W.H.S., 1907. *Malaria: a Neglected Factor in the History of Greece and Rome.* Macmillan and Bowes, Cambridge.
Kalejta-Summers, B., 1997. Diet and habitat preferences of wintering passerines on the Taff/Ely saltmarshes. *Bird Study*, **44/3**, 367–373.
Kamps, L.F., 1962. *Mud Distribution and Land Reclamation in the Eastern Wadden Shallows.* Rijkswaterstaat Communications, **4**, Den Hague, 1–73.
Kestner, F.T.J., 1962. The old coastline of the Wash. A contribution to the understanding of loose boundary processes. *Geographical Journal*, **128**, 457–478.
Kiehl, K., Eischeid, I., Gettner, S. & Walter, J., 1996. Impact of different sheep grazing intensities on saltmarsh vegetation in northern Germany, *Journal of Vegetation Science*, **7/1**, 99–106.
King, S.E. & Lester J.N., 1995. The value of saltmarsh as a sea defence. *Marine Pollution Bulletin*, **30/3**, 180–189.
Kinler, N., Linscombe, G. & Hartley, S., 1998. A *Survey of Nutria Herbivory Damage in Coastal Louisiana in 1998*. Conducted by Fur and Refuge Division Louisiana Department of Wildlife and Fisheries as part of the Nutria Harvest and Wetland Demonstration Project. Source: http://data.lacoast.gov/reports/pr/LA02PRG1.pdf.
Kittelson, P.M. & Boyd, M.J., 1997. Mechanisms of expansion for an introduced species of Cordgrass, *Spartina densiflora*, in Humboldt Bay, California. *Estuaries*, **20/4**, 770–778.
Kleyer, M., Feddersen, H. & Bockholt, R., 2003. Secondary succession on a high saltmarsh at different grazing intensities. *Journal of Coastal Conservation*, **9/2**, 123–134.
Kriwoken, L.K. & Hedge P., 2000. Exotic species and estuaries: managing *Spartina anglica* in Tasmania, Australia. *Ocean & Coastal Management*, **43/7**, 573–584.
Lacambra, C., Cutts, H., Allen, J., Burd, F. & Elliott, M., 2004. *Spartina anglica: a review of its status, dynamics and management.* Institute of Estuarine and Coastal Studies University of Hull, English Nature Research Reports, **527**. Source: http://www.english-nature.org.uk/pubs/publication/PDF/527.pdf
Laffaille, P., Lefeuvre, J.-C. & Feunteun, E., 2000. Impact of sheep grazing on juvenile sea bass, *Dicentrarchus labrax* L., in tidal saltmarshes. *Biological Conservation*, **96/3**, 271–277.
Laffaille, P., Lefeuvre, J-C., Schricke, M-T. & Feunteun, E., 2001. Feeding ecology of 0-Group Sea Bass, Dicentrarchus labrax, in saltmarshes of Mont Saint Michel (France). Estuaries, **24/1**, 116–125.
Lamberth, C. & Haycock, N., 2002. *Regulated Tidal Exchange* (**RTE**). An Intertidal Habitat Creation Technique. Report by Haycock Associates Limited, St. Albans. Environment Agency and the Royal Society for the Protection of Birds.
Laursen, K., 2002. Status of the management of geese in the Wadden Sea region in 2001. *Wadden Sea Newsletter*, **2**, Common Wadden Sea Secretariat.
Lee, W.G. & Partridge, T.R., 1983. Rates of spread of *Spartina anglica* and sediment accretion in the New River Estuary, Invercargill, New Zealand. *New Zealand Journal of Botany*, **21**, 231–236.
Leggett, D.J., Cooper, N. & Harvey, R., 2004. *Coastal and estuarine managed realignment - design issues.* Report **C628**, CIRIA. Source: http://www.ciria.org/downloads_archive.htm.
Levin, P.S., Ellis, J., Petrik, R. & Hay, M.E., 2002. Indirect effects of feral horses on estuarine communities. *The Journal of the Society for Conservation Biology*, **16/5**, 1364–1371.
Linthurst, R.A. & Reimold R.J., 1978. An evaluation of methods for estimating the Net Aerial Primary Productivity of estuarine angiosperms. *Journal of Applied Ecology*, **15/3**, 919–931.

Lipton, D.W., Wellman, K., Sheifer, I.C. & Weiher, R.F., 1995. *Economic Valuation of Natural Resources: A Guidebook for Coastal Resources Policymakers*. NOAA Coastal Ocean Program Decision Analysis Series No. **5**. Source: http://www.mdsg.umd.edu/Extension/valuation/handbook.htm

Long, S.P. & Mason, C.F., 1983. *Saltmarsh Ecology*. Blackie, Glasgow and London.

Loonen, M.J.J.E. & Bos, D., 2003. Geese in the Wadden Sea: an effect of grazing on habitat preference. In: *Challenges to the Wadden* Sea, W.J. Wolff, K. Essink, A. Kellermann & M.A. van Leeuwe, eds. 107–120. Proceedings of the 10th International Scientific Wadden Sea Symposium, Groningen, 2000. Ministry of Agriculture, Nature Management and Fisheries/ University of Groningen, Dept. of Marine Biology. Source: http://loonen.fmns.rug.nl/literatuur/iwss-loonen.pdf.

Louisiana Coastal Wetlands Conservation and Restoration Task Force, 2006. *Coastal Wetlands Planning, Protection and Restoration Act (CWPPRA): A Response to Louisiana's Land Loss*. Source: http://www.lacoast.gov/cwppra/index.htm.

Ludlam, J.P., David, H. Shull, D.H. & Buchsbaum, R., 2002. Effects of haying on salt-marsh surface invertebrates. *Biological Bulletin*, **203**, 250–251.

Lundberg, A., 1996. Changes in the vegetation and management of saltmarsh communities in southern Norway. In: *Studies in European Coastal Management*, P.S. Jones, M.G. Healy & A.T. Williams, eds., 197–206. Samara Publishing, Cardigan.

Malm, T., 1995. Excavating submerged Stone Age sites in Denmark – the Tybrind Vig example. In: *Man and Sea in the Mesolithic*, A. Fischer. ed., Oxbow Books, Oxford.

Marchant, C.J. & Goodman, P.J., 1969. *Spartina maritima* (Curtis) Fernald (in Biological Flora of the British Isles), *The Journal of Ecology*, **57/1**, 287–291.

Marchant, C.J., 1967. Evolution in *Spartina* (Gramineae). I, The history and morphology of the genus in Britain. *Journal of the Linnaean Society* (Botany), **6+A4820**, 1–26.

Marques, P.A.M. & Vicente, L., 1999. Seasonal variation of waterbird prey abundance in the Sado Estuary rice fields. *Ardeola* **46/2**, 231–234.

Martin, J.L. 2003. *The Effect of Cattle Grazing on the Abundance and Distribution of Selected Macroinvertebrates in West Galveston Island Saltmarshes*. M.Sc. Thesis, Texas A&M University. Source: https://txspace.tamu.edu/handle/1969.1/179.

Mateos Naranjo, E., Redondo Gómez, S., Luque Palomo, C.J., Castellanos Verdugo, E.M., Wharmby, C., Muñoz González, J., Palomo, M.T. & Figueroa Clemente, M.E., 2006. Analysis of the invasion of *Spartina densiflora* in the Odiel river estuary (Huelva, SW Spain). Implications for diversity. Poster presented at the 1st European Congress of Conservation Biology, 22–26 August, Eger, Hungary, Book of Abstracts, 137. Source: http://www.eccb2006.org/.

McIvor, C.C. & Rozas, L.P., 1996. Direct nekton use of intertidal saltmarsh habitat and linkage with adjacent habitats: a review from southeastern United States. In: *Estuarine Shores – Evolution, Environments and Human Activities*, K.F. Nordstrom & C.T. Roman eds., 311–334. John Wiley & Sons, New York.

McKee, K.L. & Patrick, W.H., Jr., 1988. The Relationship of Smooth Cordgrass (Spartina alterniflora) to Tidal Datums: A Review, *Estuaries*, **11/3**, 143–151.

McKee, K.L., Mendelssohn, I.A. & Materne, M.D., 2004. Acute saltmarsh dieback in the Mississippi River deltaic plain: a drought-induced phenomenon? *Global Ecology & Biogeography*, **13/1**, 65–73.

McLusky, D.S. & Elliott, M., 2004. *The Estuarine Ecosystem – Ecology, Threats and Management*, 3rd edition, Oxford University Press.

Mendessohn, I.A. & McKee, K., 1988. *Spartina alterniflora* die back in Louisiana: time course investigation of soil waterlogging effects. *Journal of Ecology*, **76**, 509–521.

Merali, Z., 2002. Secrets from Venice's Lagoon. *Sidebar, Scientific American*, Source: http://www.sciam.com/article.cfm?articleID=0008313B-E17E-1D5B-90FB809EC5880000.

Millard. A.V. & Evans. P.R., 1984. Colonisation of mudflats by *Spartina anglica* some effects on invertebrates and shore bird populations at Lindisfarne In: *Spartina anglica in Great Britain*, J.P. Doody, ed., 41–49. Focus on nature conservation, **5**, Nature Conservancy Council, Attingham Park.

Möller, I. & Spencer, T., 2002. Wave dissipation over macro-tidal saltmarshes: effects of marsh edge typology and vegetation change. *Journal of Coastal Research*, Special Issue **36**, 506–521.

Möller, I., 2003. The sea-defence role of intertidal habitats. In: *Wetland Valuation: State of the Art and Opportunities for Further Development*, L. Ledoux, ed., 73–87. Proceedings of a workshop organised for the Environment Agency by Environmental Futures Ltd and CSERGE. 19 March, 2003, University of East Anglia, Norwich, UK. Source: http://www.environment-agency.gov.uk/commondata/acrobat/workshop_proceedings_805307.pdf.

Möller, I., Spencer, T. & French, J. R., 2002. *Wave Attenuation Over Saltmarshes*. Environment Agency R&D Technical Report, W5B-022, Environment Agency.

Möller, I., Spencer, T., French, J.R., Leggett, D.J. & Dixon, M., 1999. Wave transformation over saltmarshes: a field and numerical modelling study from North Norfolk, England. *Estuarine, Coastal and Shelf Science*, **49/3**, 411–426.

Morley, J.V., 1973. Tidal immersion of *Spartina* marsh at Bridgwater Bay, Somerset. *Journal of Ecology*, **61**, 383–386.

Morris, R.K.A., Reach, I.S., Duffy, M.J., Collins, T.S. & Leafe, R.N., 2004. On the loss of saltmarshes in south-east England and the relationship with *Nereis diversicolor*. *Journal of Applied Ecology*, **41/4**, 787–791.

Moser, M., 2000. Wetland Status and Trends in Europe: the case for rehabilitation and restoration of naturally functioning wetlands. In: *The Role of Wetlands in River Basin Management, Seminar 2, Session 1: Overview of the role of wetlands in river basin functioning and assessment of wetland status and trends in Europe*, 29–36. WWF/EC Seminar – Brussels, 9–10 November 2000.

Murphy, K.C., 2005. *Progress of the 2005 Spartina Eradication Program. Report to the Legislature*. Washington State Department of Agriculture *Spartina* Program, AGR PUB 850–151 (N/1/06). Source: http://agr.wa.gov/AboutWSDA/DirectorsOffice/ReportsToLegislature.htm.

Nairn, R.G.W., 1986. *Spartina anglica* in Ireland and its potential impact on wildfowl and waders – a review. *Irish Birds*, **3**, 215–228.

National Oceanic and Atmospheric Administration (NOAA), 1991. *Coastal Wetlands of the United States: an Accounting of a National Resource Base*. National Oceanic and Atmospheric Administration, Report **91–3**.

Neckles, H.A., Dionne, M., Burdick, D.M., Roman, C.T., Buchsbaum, R. & Hutchins, E., 2002. A monitoring protocol to assess tidal restoration of saltmarshes on local and regional scales. *Restoration Ecology*, **10/3**, 556–563.

Neira, C., Levin, L.A. & Grosholz, E.D., 2005. Benthic macrofaunal communities of three sites in San Francisco Bay invaded by hybrid *Spartina*, with comparison to uninvaded habitats. *Marine Ecology Progress Series*, **292**, 111–126

Neuhaus, R., Dijkema, K.S. & Renke, H-D., 2001. The impact of sea-level rise on coastal flora and fauna. In: *Climate of the 21st Century: Changes and Consequences*, J.L., Lozán, H. GraBl, & P. Hupfer, eds., Wissenschaftliche Auswertungen, Hamburg, Germany.

Nicholls, R.J. & Hoozemans, F.M.J., 2005. Global Vulnerability Analysis. In: M. Schwartz, ed., *Encyclopaedia of Coastal Science*, 486–491. Encyclopaedia of Earth Science Series, Springer, Dordrecht, The Netherlands.

Nicholls, R.J. & Leatherman, S.P., 1995. The implications of accelerated sea level rise on developing countries. In: *Potential Impacts of Accelerated Sea-Level Rise on Developing Countries*, R.J. Nicholls & S.P. Leatherman, *Journal of Coastal Research*, Special Issue, **14**, 302–323.

Nicholls, R.J. & Wilson, T., 2001. Integrated impacts on coastal areas and river flooding, Chapter 5. In: *Regional Climate Change Impact and Response Studies in East Anglia and North West England* (RegIS), I.P. Holman & P.J. Loveland, eds., 54–103. MAFF Project No. CC0337 54. Source: http://www.silsoe.cranfield.ac.uk/iwe/projects/regis/.

Nicholls, R.J., Hoozemans, F.M.J. & Marchand, M., 1999. Increasing flood risk and wetland losses due to global sea-level rise: regional and global analyses. *Global Environmental Change*, **9**, 69–87.

# References

Niemeyer, H.D. & Kaiser, R., 2001. Hydrodynamische Wirksamkeit von Lahnungen, Hellern und Sommerdeichen. *Die Küste,* **64**, 16–58.

Normile, D., 2004. Expanding trade with China creates ecological backlash. *Science,* **306/5698**, 968–969.

Norris, K., Brindley, E., Cook, T., Babbs, S., Brown, C.F., & Yaxley, R., 1998. Is the density of redshank *Tringa totanus* nesting on saltmarshes in Great Britain declining due to changes in grazing management? *Journal Applied Ecology,* **35**/5, 621–621.

Norris, K., Cook, T., O'Dowd, B. & Durdin, C., 1997. The density of Redshank *Tringa totanus* breeding on the salt-marshes of the Wash in relation to habitat and its grazing management. *Journal of Applied Ecology,* **34/4**, 999–1013.

Nova Scotia Museum of Natural History, Undated. *The Natural History of Nova Scotia, volume 1: Topics and Habitats.* Source: http://museum.gov.ns.ca/mnh/nature/nhns/h2/h2-5.htm.

Ogburn, MB. & Alber, M., 2006. An investigation of saltmarsh dieback in Georgia using field transplants. *Estuaries and Coasts,* **29/1**, 54–62.

Oliver, F.W., 1925. *Spartina townsendii*: its modes of establishment, economic uses and taxonomic status. *Journal of Ecology,* **13**, 76–91.

Packham, J.R. & Willis, A.J., 1997. *Ecology of dunes, saltmarsh and shingle.* Chapman & Hall, London.

Paramor, O.A.L. & Hughes, R.G., 2004. The effects of bioturbation and herbivory by the polychaete *Nereis diversicolor* on loss of saltmarsh in south-east England. *Journal Applied Ecology,* **41/3**, 449–463.

Parker, B., 2005. Tides. In: *Encyclopaedia of Coastal Science,* M. Schwartz, ed., 987–995. Encyclopaedia of Earth Science Series, Springer, Dordrecht, The Netherlands.

Parlier E., Albert F., Cuzange P., Don J. & Feunteun E. In: Press. The 'herbus' of Mont Saint-Michel Bay and the 'mizottes' of Aiguillon Bay: Impact of human disturbance on the nursery function of tidal marshes. *Cahier de biologie marine.* Source: http://www.univ-lr.fr/Labo/lbem/pdf/Parlier_CV.pdf.

Patridge, T.R., 1987. *Spartina* in New Zealand. *New Zealand Journal of Botany,* **25**, 567–575.

Patten, K., 2002. Smooth Cordgrass (*Spartina alterniflora*) Control with Imazapyr. *Weed Technology,* **16/4**, 826–832.

Patterson, I.J., 1991. Conflict between geese and agriculture – does goose grazing cause damage to crops. *Ardea,* **79/2**, 178–186.

Pearce, D. & Barbier, E.B., 2000. *Blueprint for a Sustainable Economy.* The Blueprint Series, **6**. Earthscan Publications Ltd.

Pearce, D., Markandya, A. & Barbier, E.B., 1989. *Blueprint for a Sustainable Economy.* The Blueprint Series, **1**. Earthscan Publications Ltd.

Percival, S.M., Sutherland, W.J. & Evans, P.R., 1998. Intertidal habitat loss and wildfowl numbers: applications of a spatial depletion model. *Journal of Applied Ecology,* **35**, 57–63.

Pethick, J., 1984. *An introduction to coastal geomorphology,* Edward Arnold, London.

Pethick, J.S., 2001. Coastal management and sea level rise. *Catena,* **42**, 307–322.

Pétillon, J, Ysnel, F, Lefeuvre, J-C, & Canard, A., 2005. Are saltmarsh invasions by the grass *Elymus athericus* a threat for two dominant halophytic wolf spiders? *Journal of Arachnology,* **33/2**, 236–242.

Pilkey, O.H. & Cooper, J.A.G., 2004. Society and Sea Level Rise. *Science,* **303/5665**, 1781–1782.

Portela, L.I., 2002. Preliminary assessment of saltmarsh areas in the Tagus estuary. In: *Littoral, 2002, The Changing Coast. EUROCOAST / EUCC, Porto – Portugal,* 117–119.

Postma, H., 1961. Mechanism of mud accumulation in estuaries: Transport and accumulation of suspended matter in the Dutch Wadden Sea. *Netherlands Journal of Sea Research.* **1**, 148–190.

Preston, C.D., Pearman, D.A. & Dines, T.D., 2002. *New Atlas of the British & Irish Flora.* Oxford University Press.

Pringle, A.W., 1995. Erosion of a cyclic saltmarsh in Morecambe Bay, north-west England. *Earth Surface Processes and Landforms,* **20/5**, 387–405.

Pye, K. & French, P.W., 1992. *Targets for coastal habitat re-creation*. Unpublished report English Nature Science, **13**, Peterborough (SC1.13).

Pye, K. & French, P.W., 1993. *Erosion & Accretion Processes on British Saltmarshes*. Cambridge Environmental Research Consultants Ltd, Cambridge, 5 volumes.

Pye, K., 1996. Evolution of the Dee estuary, United Kingdom. In: *Estuarine shores: evolution, environments and human alterations*, K.F. Nordstrom & C.T. Roman, eds., 15–37. John Wiley & Sons.

Qin, P., Xie, M., Jiang, Y. & Chung, C-H., 1997. Estimation of the ecological-economic benefits of two *Spartina alterniflora* plantations in North Jiangsu, China. *Ecological Engineering*, **8/1**, 5–17.

Ranwell, D.S., 1964. *Spartina* saltmarshes in southern England. II. Rate and seasonal pattern of sediment accretion. *Journal of Ecology*, **52**, 79–95.

Ranwell, D.S., 1967. World resources of *Spartina townsendii* (sensu lato and economic use of *Spartina* marshland. *Journal of Applied Ecology*, **4**, 239–256.

Ranwell, D.S., 1972. *Ecology of saltmarshes and sand dunes*. London, Chapman & Hall, 258 pp.

Ratcliffe, D.A., 1977. *A Nature Conservation Review: the Selection of Biological Sites of National Importance to Nature Conservation in Britain*. Vols. 1 & 2, Cambridge University Press, Cambridge.

Reise, K., 2005. Coast of change: habitat loss and transformations in the Wadden Sea. *Helgoland Marine Research*, **59/1**, 9–21.

Roach, E.R., Watzin, M.C. & Scurry, J.D., 1987. Wetland changes in coastal Alabama. In: *Symposium on the Natural Resources of the Mobile Bay Estuary*, T.A. Lowery, ed. Alabama Sea Grant Extension Service, Mobile, **AL. MASGP-87-007**, 92–101.

Roberts, P.D. & Pullin, A.S., 2006. The effectiveness of management options used for the control of *Spartina* species. *Systematic Review*, **22**, Centre for Evidence-Based Conservation, Birmingham, UK.

Robinson, D., 1987. The Wash: geographical and historical perspectives. In: *The Wash and Its Environment*, J.P. Doody & B. Barnett, eds., 23–34. Report of a Conference held on 8–10 April, 1987 at Horncastle, Lincolnshire.

Robinson, N.A., 1984. The history of Spartina in the Ribble Estuary. In: *Spartina anglica in Great Britain*, J.P. Doody, ed., 27–29. Report of a meeting held at Liverpool University, 10th November 1982. Nature Conservancy Council, Attingham Park.

Rodwell, J.S., ed., 2000. *British Plant Communities. Volume 5, Maritime Communities and Vegetation of Open Habitats*. Cambridge University Press, Cambridge.

Rogers, J.N., 2005. *Saltmarsh Assessment and Restoration Tool (SMART): An Evaluation of Hyperspectral, LIDAR and SWMM for Producing Accurate Habitat Restoration Predictions*. Cooperative Institute for Coastal and Estuarine Environmental Technology Project. Source: http://ciceet.unh.edu/news/releases/smartRelease.html.

Roman, C.T. & Daiber, F.C., 1984. Aboveground and Belowground Primary Production Dynamics of Two Delaware Bay Tidal Marshes. *Bulletin of the Torrey Botanical Club*, **111/1**, 34–41.

Roman, C.T., Niering, W.A. & Warren, R.S., 1984. Saltmarsh vegetation change in response to tidal restriction. *Environmental Management*, **8/2**, 141–149.

Rupp, S. & Nicholls, R.J., 2002. Managed Realignment of Coastal Flood Defences: A Comparison between England and Germany. In: Proceedings of *Dealing with Flood Risk* An interdisciplinary seminar of the regional implications of modern flood management, B. van Kappel ed., Delft Hydraulics. Source: http://www.survas.mdx.ac.uk/pdfs/delft_pa.pdf.

Ruxton, T.D. & Baker, A.C.J., 1978. Engineering aspects of estuarial water storage schemes in the U.K. *Aquatic Ecology*, **12/3–4**, 277–290.

Sadoul, N., Walmsley, J.G. & Charpentier, B., 1998. *Salinas and Nature Conservation*. Conservation of Mediterranean Wetlands, **9**, Tour du Valat, Arles (France).

San Francisco Estuary Project, 2000. *State of the Estuary Report 2000, Restoration Primer*. San Francisco Estuary Program. Source: http://sfep.abag.ca.gov/pdf/restoration_primer/Restoration_Primer%5Bp1-15%5D.pdf

# References

San Francisco Estuary Project, 2005. *State of the Estuary Report 2004*, San Francisco Estuary Program. Source: http://sfep.abag.ca.gov/pdf/restoration_primer/

Sánchez J.M., SanLeon D.G. & Izco J., 2001. Primary colonisation of mudflat estuaries by *Spartina maritima* (Curtis) Fernald in Northwest Spain: vegetation structure and sediment accretion. *Aquatic Botany*, **69/1**, 15–25.

Sayce, K., 1988. *Introduced cordgrass, Spartina alterniflora Loisel., in saltmarshes and tidelands of Willapa Bay*, Washington. Report for Willapa National Wildlife Refuge, Washington, US Fish and Wildlife Service, Contract FWSI–87058. Source: http://www.friendsofwillaparefuge.org/cordgrassstudy1.pdf.

Schröder, H.K., Kiehl, K. & Stock, M., 2002. Directional and non-directional vegetation changes in a temperate saltmarsh in relation to biotic and abiotic factors. *Applied Vegetation Science*, **5/1**, 33–44.

Schwimmer, R.A. & Pizzuto, J.E., 2000. A model for the evolution of marsh shorelines. *Journal of Sedimentary Research*, **70**, 1026–1035.

Scott, R., Callaghan, T.V. & Lawson, GJ., 1990. *Spartina* as a biofuel. In: A.J. Gray & P.E.M. Benham, eds. *Spartina anglica – A research review*. Institute of Terrestrial Ecology, Research publication No. **2**, HMSO, 48–51.

Shaw, W.B. & Gosling D.S., 1997. *Spartina* ecology, control and eradication – recent New Zealand experience. In: *Second International Spartina Conference*: Proceedings, Olympia WA, 27–33.

Shennan, I & Horton, B., 2002. Holocene land- and sea-level changes in Great Britain. *Journal of Quaternary Science*, **17/5–6**, 511–526.

Shennan, I., 1987. Impacts on the Wash of sea level rise. In: *The Wash and its environment*, J.P. Doody & B. Barnet, eds., 77–90. Research and survey in nature conservation, **7**, Nature Conservancy Council, Peterborough.

Silliman, B.R. & Bertness, J.C. 2002. Top-down control of *Spartina alterniflora* production by Periwinkle grazing, *PNAS*, **99/16**, 10500–10505

Silliman, B.R. & Zieman, J.C. 2001. Top-down control of *Spartina alterniflora* production by Periwinkle grazing in a Virginia saltmarsh. *Ecology*, **82/10**, 2830–2845.

Silvestri, S., Marani, M. & Marani, A., 2003. Hyperspectral remote sensing of saltmarsh vegetation, morphology and soil topography. *Physics and Chemistry of the Earth*, **28**, 15–25.

Sir William Halcrow and Partners Ltd, 1988a. *The sea defence management study for the Anglian Region*. Strategy report for Anglian Water.

Sir William Halcrow and Partners Ltd, 1988b. *The Anglian Coastal Management Atlas*. Prepared for Anglian Water.

Stade Declaration, 1998. *Trilateral Wadden Sea Plan*. Ministerial declaration of the 8th trilateral governmental conference on the protection of the Wadden Sea. Common Wadden Sea Secretariat, Wilhelmshaven, Germany. Source: http://www.waddensea-secretariat.org/tgc/MD-Stade.html.

Stapf, O., 1908. *Spartina townsendii*. *Journal of Botany*, **46**, 76–81.

Stewart, R.E. Jr., Proffitt, C.E. & Charron, T.M., 2001. *Abstracts from 'Coastal Marsh Dieback in the Northern Gulf of Mexico: Extent, Causes, Consequences, and Remedies'*. Information and Technology Report USGS/BRD/ITR—2001-0003, US Department of the Interior, US Geological Survey. Source: http://brownmarsh.net/reports/abstracts.pdf.

Stock, M. & Hofeditz, F., 2003. Impact of sheep grazing on habitat utilisation of Barnacle Geese (*Branta leucopsis*) on saltmarshes - implications for management. In: *Challenges to the Wadden Sea*, W.J., Wolff, K. Essink, A. Kellermann & M.A. van Leeuwe, eds., 89–106. Proceedings 10th International Scientific Wadden Sea Symposium, Groningen. Ministry of Agriculture, Nature Management and Fisheries, The Hague and Fisheries and the Department of Marine Biology, University of Groningen.

Strachan, D., 1998. *Essex from the Air*. Essex County Council.

Stralberg, D., Toniolo, V., Page, G.W. & Stenzel, L.E., 2004. *Potential Impacts of Non-Native Spartina Spread on Shorebird Populations in South San Francisco Bay*. PRBO Report to California Coastal Conservancy (contract 02–212). Point Reyes Bird Observatory Conservation Science, Stinson Beach, CA. Source: http://www.prbo.org/wetlands/prbo_spartina_feb04.pdf.

Streever, W.J., 2000. *Spartina alterniflora* marshes on dredged material: a critical review of the ongoing debate over success. *Wetlands Ecology and Management*, **8/5**, 295–316.

Strong, D.R. & Ayres, D.R., 2005. *Dynamics and Ecosystem threats of Bidirectional Cordgrass Hybridization in San Francisco Bay*. California Sea Grant College Program, Research Completion Reports, Paper Coastal05 01. Source:

Taylor, J.A., Murdock, A.P. & Pontee, N.I., 2004. A macroscale analysis of coastal steepening around the coast of England and Wales. *The Geographical Journal*, **170/3**, 179–188.

Taylor, M.C. & Burrows, E.M., 1968. Studies in the biology of *Spartina* in the Dee estuary, Cheshire. *Journal of Ecology*, **56**, 795–809.

Teal, J. & Teal, M., 1969. *Life and death of the saltmarsh*. Ballantine Books, New York.

Teal, J.M. & Weinstein, M.P., 2002. Ecological engineering, design, and construction considerations for marsh restorations in Delaware Bay, USA. *Ecological Engineering*, **18**, 607–618.

Tessier, M., Vivierb, J-P., Ouinc, A., Gloaguenb, J-C. & Lefeuvreb, J-C., 2003. Vegetation dynamics and plant species interactions under grazed and ungrazed conditions in a western European saltmarsh. *Acta Oecologica*, **24/2**, 103–111.

Thom, R.M., 1992. Accretion rates of low intertidal saltmarshes in the Pacific Northwest. *Wetlands* **12/2**, 147–156.

Thompson, J.D., 1991. The Biology of an Invasive Plant. *BioScience*, **41/6**, 393–401.

Titus, J.G., 1991. Greenhouse Effect And Coastal Wetland Policy: How Americans Could Abandon An Area The Size Of Massachusetts At Minimum Cost. *Environmental Management*, **15/1**, 39–58.

Toft, A.R., Pethick, J.S., Burd, F., Gray, A.J., Doody, J.P. & Penning-Rowsell, E., 1995. *A guide to understanding and management of saltmarshes*. National Rivers Authority, Research & Development Note **324**, Foundation for Water Research, Marlow.

Trinkley, M. & Fick, S., 2003. *Rice Cultivation, Processing, and Marketing in the Eighteenth Century*. Chicora Foundation, Inc., Columbia. [Liberty Hall (Research Series 62) data recovery project]. Source: http://www.chicora.org/Rice%20Context.pdf.

Truscott, A., 1984. Control of *Spartina anglica* on the amenity beaches of Southport. In: *Spartina anglica in Great Britain*, J.P. Doody, ed., 64–69. Focus on nature conservation, **5**, Nature Conservancy Council, Attingham Park.

Tubbs, C., 1984. *Spartina* on the south coast: an introduction. In: *Spartina anglica in Great Britain*, J.P Doody, ed., 3–4. Focus on nature conservation, **5**, Nature Conservancy Council, Attingham Park,.

Tucker, G.M. & Evans, M.I., eds., 1997. Habitats for Birds in Europe. A Conservation Strategy for the Wider Environment. Birdlife Conservation Series **6**, Birdlife International.

Turner, R.E. & Streever, B., 2002. *Approaches to Coastal Wetland Restoration: Northern Gulf of Mexico*. SPB Academic Publishing, The Hague, Netherlands.

Ungar, I.A., 1998. Are Biotic Factors Significant in Influencing the Distribution of Halophytes in Saline Habitats? *The Botanical Review*, **64/2**, 176–199.

US Army Corps of Engineers, 1987. Engineering and Design - Beneficial Uses of Dredged Material. Engineer Manual, No. 1110-2-5026. Source: http://www.usace.army.mil/publications.

US Army Corps of Engineers, 2006. Louisiana Coastal Protection and Restoration, Preliminary Technical Report. Source: http://lacpr.usace.army.mil/default.aspx?p=report.

van de Koppel, J., van der Wal, D., Bakker, J.P. & Herman, P.M.J., 2005. Self-organisation and vegetation collapse in saltmarsh ecosystems. *The American Naturalist*, **165/1**, 1–12.

van der Graaf, A.J., Bos, D., Loonen, M.J.J.E., Engelmoer, M. & Drent, R.H., 2002. Short-term and long-term facilitation of goose grazing by livestock in the Dutch Wadden Sea area. *Journal of Coastal Conservation*, **8/2**, 179–188.

van der Wal, D., Pye, K, & Neal, A., 2002. Long-term morphological change in the Ribble Estuary, northwest England. *Marine Geology*, **189**, 249–266.

Van der Wal, R., Van Wijnen, H., Van Wieren, S., Beucher, O. & Bos, D., 2000. On Facilitation Between Herbivores: How Brent Geese Profit From Brown Hares. *Ecology*, **81/4**, 969–980.

Van Eerden, M.R., Drent, R.H., Stahl, J. & Bakker, J.P., 2005. Connecting seas: western Palaearctic continental flyway for water birds in the perspective of changing land use and climate. *Global Change Biology*, **11/6**, 894–908.

van Wijnen, H.J, Bakker, J.P. & Vries, Y., 1996. Twenty years of saltmarsh succession on a Dutch barrier island. *Journal of Coastal Conservation*, **3/1**, 9–18.

Vickery, J.A., Watkinson, A.R. & Sutherland, W.J. 1994. *The Solutions to the Brent Goose Problem: An Economic Analysis*. The Journal of Applied Ecology, **31/2**, 371–382.

Viñals, M.J., Bernués, M., Dugan, P, Llopart, P. & T. Salathé, 2001. *Ramsar Advisory Missions*, 43, Ebro Delta, Catalonia Spain. Source: http://www.ramsar.org/ram/ram_rpt_43e_summ.htm.

Vinther, N., Christiansen, C. & Bartholdy, J., 2001. Colonisation of *Spartina* on a tidal water divide, Danish Wadden Sea. *Geografisk Tidsskrift. Danish Journal of Geography*, **101**, 11–20.

Walmsley, J.G., 1999. The Ecological importance of Mediterranean salinas. 6th Conference on Environmental Science and Technology, Dept. of Environmental Studies University of the Aegean, Pythagorion, Samos, Greece 30th August – 2nd September 1999. Post Conference Symposium on *'Saltworks: Preserving Saline Coastal Ecosystems'*, 81–95.

Wang, Y.Q. & Christiano, M., 2006. *A Protocol of Using High Spatial Resolution Satellite Imagery to Map Saltmarshes: Towards Long Term Change Detection*. A final report of a cooperative project by the Laboratory for Terrestrial Remote Sensing (LTRS), the Department of Natural Resources Science, University of Rhode Island and the National Park Service. Source: http://www.ltrs.uri.edu/research/GatewayFinalReport07282006.pdf.

Warren, G.M., 1911. *Tidal marshes and their reclamation*. US Department of Agriculture, Experimental Station Bulletin, 240.

Warren, R.S., Fell, P.E., Rozsa, R., Brawley, A.H., Orsted, A.C., Olson, E.t., Swamy, V. & Niering, W.A., 2002. Saltmarsh restoration in Connecticut: 20 years of science and management. *Restoration Ecology*, **10/3**, 497–513.

Webb, J.W. & Dodd, J.D., 1983. Wave-protected versus unprotected transplantings on a Texas bay shoreline. *Journal of Soil and Water Conservation*, **38/4**, 363–366.

Webb, J.W. & Dodd, J.D., 1989. *Spartina alterniflora* response to fertilizer, planting dates, and elevation in Galveston Bay, Texas. *Wetlands*, **9/1**, 61–72.

Wheelwright, N. T. & Rising, J. D., 1993. Savannah Sparrow (*Passerculus sandwichensis*). In: *The Birds of North America*, A. Poole and F. Gill eds., 45. The Academy of Natural Sciences, Philadelphia; the American Ornithologists' Union Washington, DC.

White, W.A., Tremblay, T.A. Wermund, E.G. & Handley L.R.. 1993. *Trends and Status of Wetland and Aquatic Habitats in the Galveston Bay System, Texas*. The Galveston Bay National Estuary Program. GBNEP, **31**, 225 pp.

Whiteside, M.R., 1987. *Spartina* colonisation. In: *Morecambe Bay, an Assessment of Present Ecological* Knowledge, N.A. Robinson & A.W. Pringle, eds., 118–129. Morecambe Bay Study Group with the University of Lancaster.

Wikipedia, 2007. *Severn Barrage*. Source: http://en.wikipedia.org/wiki/Severn_Barrage#Thomas_Fulljames_-_1849.

Williams, A.T. & Phillips, M.R., 2005. Environmental risk assessment – a current synopsis of the Cardiff Bay Barrage, Wales, UK. In: *International Conference on Coastal Conservation and Management in the Atlantic and Mediterranean*, A. Sena, O. Ferreira, P. Noronha, F. Veloso Gomes, F. Taveira Pinto, F. Correira, & L. das Neves, eds., 47–50. Tavira, Portugal, 17 – 20 April, 2005.

Williams, L., Noblitt G.C. & Buchsbaum, R., 2001. The Effects of Saltmarsh Haying on Benthic Algal Biomass, *Biological Bulletin*, **201**, 287–288.

Williams, P.B. & Orr, M.K., 2002. Physical Evolution of Restored Breached Levee Saltmarshes in the San Francisco Bay Estuary. *Restoration Ecology*, **10/3**, 527–527.

Williamson, R. 1995. *Spartina* in Victoria: an overview. In: *Proceedings of the Australasian Conference on Spartina Control*, J.E. Rash, R.C. Williamson & S.J. Taylor, eds., Victorian Government Publication, Melbourne, Australia.

Winn, P.J.S., Young, R.M. & Edwards, A.M.C. 2003. Planning for the rising tides: the Humber Estuary Shoreline Management Plan. *The Science of the Total Environment*, **314–316**, 13–30.

Wolfram, Ch., Hörcher, U., Lorenzen, D., Neuhaus, R., Aegerter, E. & Dierßen, K. 1998. 1. *Vegetation succession in a salt-water lagoon in the polder Beltringharder Koog, German Wadden Sea*. Proceedings of the yearly IAVS Symposium – 2000, **1**, 43–46.

Wolters, M., Garbutt, A. & Bakker, J.P., 2005. Salt-marsh restoration: evaluating the success of de-embankments in north-west Europe. *Biological Conservation*, **123**, 249–268.
Wong, P.P., 2005. Reclamation. In: *Encyclopedia of Coastal Science*, M.L. Schwartz ed., 791–794. Encyclopaedia of Earth Science Series, Springer, Dordrecht, The Netherlands.
Worrall, S., 2005. The UK LIFE project on shoreline management: 'Living with the Sea'. In: J.-L., Herrier J. Mees, A. Salman, J. Seys, H. Van Nieuwenhuyse & I. Dobbelaere, eds. *Proceedings 'Dunes and Estuaries 2005' - International Conference on Nature Restoration Practices in European Coastal Habitats*, Koksijde, Belgium, 19–23 September, 2005, 451–459. VLIZ Special Publication, **19**.
Wu, M-Y., Hacker, S., Ayres, D. & Strong, D.R., 1999. Potential of *Prokelisia* spp. as Biological Control Agents of English Cordgrass, *Spartina anglica*. *Biological Control*, **16/3**, 267–273.
Yan, X., Zhenyu, L., Gregg, W.P., & Dianmo, L., 2000. Invasive Species in China – An Overview. *Biodiversity and Conservation*, **10/8**, 1317–1341.
Yapp, R.H., Johns, D. & Jones, O.T., 1917. The saltmarshes of the Dovey Estuary. II, The saltmarshes. *Journal of Ecology*, **5**, 65–103.
Zedler, J.B., 2000. *Handbook for Restoring Tidal Wetlands*. CRC Press.
Zhang, R.S., Shen, Y.M., Lu, L.Y., Yan, S.G., Wang, Y.H., Li, J.L. & Zhang, Z.L., 2004. Formation of *Spartina alterniflora* saltmarshes on the coast of Jiangsu Province, China. *Ecological Engineering*, **23/2**, 95–105.

# Index

**A**

Accreting saltmarshes, summary definition, 52
Accretion, 2, 5, 37, 40, 42, 43, 50, 51, 54, 55, 76, 81–83, 87, 88, 143, 167, 184
Accretion rates, Severn Estuary, 5
Albania, 15, 25, 26, 34, 68
Alkborough realignment, Humber Estuary, 104, 169, 179
Amenity barrage, Cardiff Bay, 41
Amphidromic point, 4
Arctic Goose Habitat Working Group, 122
Avian herbivores, livestock grazing, trends and trade-offs, 116–118

**B**

Bay of Fundy, 4, 81, 175
Beltringharder, German Wadden Sea, 92, 173
Black Rail, 72
Blakeney, North Norfolk, 63
Bridgwater Bay, 18, 143, 158, 162

**C**

Cardiff Bay barrage, 35, 41, 44
Coastal cells, 177, 178
Coastal wetland loss
   Gulf of Mexico, 29, 82
   Louisiana, 82, 174
Coastal Wetlands Planning, Protection and Restoration Act, 174
Colonisation, *Spartina*, 29, 51, 87, 139–141, 143, 155
Cudmore Grove, polder, 42, 43
Cultural values
   archaeological, 64
   landscape, 63
   nature conservation, 64
   recreation, 63
   research and teaching, 63

**D**

Dee Estuary, 5, 26, 78, 148
Delta Project, 29
Delta region, 177
Dengie peninsula, 40, 60, 96
Die-back, *Spartina*, 97, 98, 154–157
Driving forces and pressures
   accretion, 51
   erosion, 56
   natural dynamics, 54
Dudden Estuary, Cumbria, 66
Dynamic equilibrium, summary definition, 55

**E**

Eatouts by the muskrat, 110
Ebro Delta, 23, 25, 180, 181
Economic values
   commercial fish, 61
   grazing by domestic stock, 78
   improved water quality, 61, 62
   Sea Bass, 61
   sea defence, 78
Ecosystem values, 58, 78
   recycling mechanisms, 58
   specialised plants and animals, 58
   trophic energy, 58
Enclosure
   for agricultural, consequences, 37, 38
   cessation of in the Wadden Sea, 113
   forms of, 20
   grazing marsh, 113, 126
   of tidal lagoons, creation of salinas, 22
Eroding saltmarsh, summary definition, 79, 80
Erosion
   lateral, 56

River Deben, Suffolk, 56
slumping, 54, 56, 79, 155
Essex
    land at risk from flooding, 100, 182
    loss of saltmarsh, 40–42
    Northey Island realignment, 38–40
Estuaries review UK, 36
Estuarine habitat loss, USA, 33
Estuary Restoration Act, USA, 33, 95
EU Demonstration Programme on Integrated Coastal Zone Management, 46
European Coastal Charter, 45
European Parliament, 46
European saltmarsh inventory, 34
European Union erosion project, 104
Eustasy, 51, 54, 56, 166

**F**
Fenland Basin, the coastal squeeze, 169
Flood of 1953, 100, 177
Food web, 10, 58, 111, 114
Freiston Shore, realignment, 38, 44

**G**
Galveston Bay, 33, 82, 94, 95
German Baltic coast, saltmarsh restoration, 99
Global sea level change, 166
Global warming, 31, 40, 166, 169, 170, 173, 179, 186
Grazing
    changes in vegetation, 118
    controlling wintering ducks and geese, 17, 61, 71, 123
    domestic livestock, 112
    by domestic stock, moderate, 61, 132
    effects on wintering passerines, 71
    invertebrates, 118, 119
    native herbivores, 120
    native mammals, 110
    Nutria, 110–111
    and plant species diversity, 114, 123, 137
    and rare plants, trends and trade-offs, 112–114
    State Evaluation Model, 123–126, 131, 133, 134
    wintering of ducks and geese, 17
Grazing regimes, reasons for changing, 69, 138
Great Fen Project, 170
Gulf of Mexico, use by fish and macro crustaceans, 61

**H**
Historically, ungrazed saltmarshes, 132
Horizontal saltmarsh growth rates, 50
Hurricane Katrina, 174
Hybridisation, *Spartina*, 141, 142, 163

**I**
Infrastructure
    conservation case, 35, 36
    Delta works, the Netherlands, 35
    projects UK, 35
    saltmarsh loss, 24, 30
Invasion, *Spartina*, 51, 85, 163
Invertebrates
    distribution across a saltmarsh, 118
    grazing trends and trade-offs, 118, 119
Isostasy, 51, 54, 56
Isostatic post-glacial rebound, 166

**L**
Land claim, the Wash, eastern England, 20, 27, 36, 62, 117
Legislation and realignment sites, UK, 104, 185
Lesser Snow Goose, range, 110, 120
Lindisfarne National Nature Reserve, 148, 151, 152, 158
Livestock grazing, breeding birds, trends and trade-offs, 116, 117
Livestock removal, trends in plant communities, 112–114
Living with the sea, an historical perspective, 183, 184
Living with the Sea, project, 46, 104
Loch-head saltmarsh, 7, 14
Loss of amenity values, Dee Estuary, 148, 149
Loss of saltmarsh, Essex coast, 38, 41, 42, 184
Louisiana coastal protection and restoration, 174

**M**
Macro-tidal, 7, 60, 75, 77
Malaria, 25
Managed realignment
    Northey Island, Essex, 38–40
    opportunities, 168–170, 178
Mediterranean, micro-tidal deltas, 11
Mediterranean deltas
    mosquitoes and malaria, 25
    rice cultivation, value for birds, 23
Meso-tidal, 7

Micro-tidal, 7, 11, 180
Mississippi Delta, 174, 175, 180, 181
Morecambe Bay, 20, 52
    European Marine Site, 138
    saltmarsh erosion, 52
    SSSI, 129
    turf cutting, 20
Mosquitoes, 25, 99
Mowing, 135, 137, 152, 153

## N

Natura 2000, vegetation types, 46, 47
Natural England, 20, 38, 96, 107, 129, 157, 162
Naturally restored saltmarsh, R. Deben, Suffolk, 56, 102
Net primary productivity, 8, 9
Nutria/Coypu damage, 110, 111

## O

Odiel Marshes, 11
Over-grazing, geese and saltmarsh vegetation, 56, 64, 65, 121–123, 126

## P

Physical and biological interactions, 93
Physical and dynamic factors promoting, erosion, 76
Physical and hydrodynamic factors promoting accretion, 76
Physical values
    trends in accreting saltmarshes, 50–52
    trends in dynamically stable saltmarsh, 52–55
    trends in eroding saltmarsh, 55–57
Piel Channel Flats, Morecambe Bay, *Spartina* expansion, 160
Plant communities, nature conservation values, 64, 69
Plant establishment, desirable conditions, 94
Prehistoric people
    reed cutting, 18, 19
    Severn Estuary, 18
Primary productivity, 7–9, 61, 85, 93, 95

## R

Ragworms and saltmarsh development - a debate, 112
Rare plants of saltmarsh, 127
Realignment
    different approaches in the Wadden Sea, 172, 173
    engineering requirements, 103, 104
Reclamation, use of word, 30
Re-creating soil characteristics, timing, 93
Recreational activity, 31, 78
Redshank, affects of grazing on breeding, 116, 117
Reed use, American Indians, 19
Reference conditions, 84
Regulated tidal exchange, 99, 103, 172, 173
Reintroduction of grazing, 126, 135, 137
Relative sea level change, 89, 166
Remote sensing, 83, 84
Restoration
    promoting accretion, 88
    protecting existing saltmarsh, 88
    re-integration, 89
Restoration methods
    beneficial use of dredged materials, 93, 94
    offshore breakwaters, 95–97
    rip-rap, 97
    *Spartina*, 94, 95
    vegetation, 94
    warping, 91, 92
Restore America's Wetlands, 95, 175
Restoring saltmarsh
    abandonment, 100–102
    realignment, 102–106
Restoring tidal restricted flows, 98, 99
Rias, 11
Ribble Estuary, 78, 116, 149, 191

## S

Sado Estuary, 15
Salinas
    importance for birds, 22, 23
    traditional, 22, 23
Saltmarsh and sea level rise, 6, 7, 31, 37, 43, 55, 56, 77, 79, 83, 88, 112
Saltmarsh conservation, changing attitudes, 185
Saltmarsh distribution
    area, 11, 40, 44, 52, 83
    Europe, 12
    Great Britain, 11, 13, 30, 71, 115, 157
Saltmarsh hay, 17, 19, 119, 131
Saltmarsh Lamb, Pré-Salé, 60
Saltmarsh loss
    land claim, 24
    world-wide, 30, 83
Saltmarsh plants

response to abandonment of grazing, 67, 68
response to heavy grazing, 65, 66
response to little or no grazing, 66, 67
response to moderate grazing, 66
Saltmarsh restoration
 and monitoring-web sites, 106, 107
 southern North Sea, 168, 169
Saltmarsh states
 physical, 75–77
 vegetative, 64–69
Saltmarsh values, 64, 89, 126, 131
 physical state, 75–77
 vegetative, 64
Saltmarsh vegetation
 Mediterranean, 7, 11, 67, 70, 179, 180
 North America, 11, 14, 19, 60, 65, 70, 131
 temperate, 6, 12, 51, 55, 110
Saltmarshes and sea level rise, 6, 31, 75, 77, 79, 82, 83, 167, 168
Saltmarshes of North America, vegetation types, 70
Saltpans, 6, 53–55, 79
Samphire, 17, 19, 62, 77, 79, 80
Schleswick-Holstein, 91, 92
Sediment accumulation, factors affecting, 7
Sediment sources, 1, 2, 177
Sedimentation rates, *Spartina*, 143
Setbacks, *Spartina* restoration, 97, 98
Severn Bore, 4
Severn Estuary, 5, 11, 18, 19, 97
 Cardiff Bay, 41
Siltation, Ribble Estuary, 78
Solway Firth, Cumbria, 67
South Walney Island, 158–160
*Spartina*
 changing patterns of invasion, 157, 158
 die-back, 155–157
 export of plants, 144
 impacts on amenity beaches, 148, 149
 impacts on bird populations, 148
 patterns of invasion, Bridgwater Bay, 158
 patterns of invasion, Lindisfarne, 157, 158
 succession in northwest England, 158–161
*Spartina* accretion rates, 51, 150
*Spartina alterniflora,* problems in the USA, 149, 150
*Spartina anglica,* as a 'native' species, 145, 146, 163
*Spartina* control
 biological, 152, 153
 costs and concerns, 153
 grazing, 152
 herbicides, 151, 152
 physical/mechanical methods, 152
*Spartina densiflora,* expansion in the Odiel Estuary, 163, 164
*Spartina* hybridisation, 141, 142, 163
*Spartina maritima* clones, 141
*Spartina*, native and introduced species
 in North America, 15, 70, 95, 97, 98, 139, 141, 142, 147, 154
*Spartina* spread in China, Australia or New Zealand, 145
Special Area of Conservation, Wash and North Norfolk, 128, 129
State Evaluation Model
 physical, 81
 vegetative, 124, 125
Stocking rates
 dominance of *Elytrigia atherica*, 1, 29, 130, 135
 species trends and trade-offs, 114–116

**T**
Tagus estuary, 85
The Wadden Sea, realignment, 173, 176
The Wash
 accreting saltmarsh, 28, 36
 embankment, 26, 27, 37
 nature conservation case against enclosure, 35, 36
Tidal creeks, 30, 66, 80
Tidal cycle, 4, 173
Tidal range, 4, 7, 35, 74, 89, 131, 139, 142, 144, 173
Tide tables, 4
Tollesbury realignment, Blackwater Estuary, 104
Trends and trade-offs
 dynamic equilibrium, 78, 79
 eroding saltmarshes, 79, 80
Trends and trades-offs, accreting saltmarsh, 77, 78

**V**
Vegetation
 geographic scales, 6
 succession, 1, 4, 5, 19, 20, 50, 136, 173
Vegetation management, reducing grazing pressure, 134, 135
Vegetation patterns, 6, 50, 85, 108
Vegetation values
 breeding passerines, 72
 breeding waterfowl, 71, 72

invertebrates, 72, 73
mammal populations, 73
wintering passerines, 71
wintering wildfowl, 71
Venice lagoon, 11, 180, 181
Vermuyden, Cornelius Sir, 26
Vertical growth rates, 51, 52

**W**
Wadden Sea, warping and sediment fields, 28, 29

Wash Site of Special Scientific Interest, 129
Wave attenuation, 58–60

**Y**
Yzer River mouth, Belgium
realignment, 171, 172

**Z**
Zuiderzee, 29

# Coastal Systems and Continental Margins

1. B.U. Haq (ed.): *Sequence Stratigraphy and Depositional Response to Eustatic, Tectonic and Climatic Forcing.* 1995　　　　　　　　ISBN 0-7923-3780-8
2. J.D. Milliman and B.U. Haq (eds.): *Sea-Level Rise and Coastal Subsidence. Causes, Consequences, and Strategies.* 1996　　　　　　　　ISBN 0-7923-3933-9
3. B.U. Haq, S.M. Haq, G. Kullenberg and J.H. Stel (eds.): *Coastal Zone Management Imperative for Maritime Developing Nations.* 1997　　　　ISBN 0-7923-4765-X
4. D.R. Green and S.D. King (eds.): *Coastal and Marine GeoInformation Systems: Applying the Technology to the Environment.* 2001　　　ISBN 0-7923-5686-1
5. M.D. Max (ed.): *Natural Gas Hydrate in Oceanic and Permafrost Environments.* 2000
　　　　　　　　　　　　　　　　　　ISBN 0-7923-6606-9; Pb 1-4020-1362-0
6. J. Chen, D. Eisma, K. Hotta and H.J. Walker (eds.): *Engineered Coasts.* 2002
　　　　　　　　　　　　　　　　　　　　　　　ISBN 1-4020-0521-0
7. C. Goudas, G. Katsiaris, V. May and T. Karambas (eds.): *Soft Shore Protection. An Environmental Innovation in Coastal Engineering.* 2003　　　ISBN 1-4020-1153-9
8. D.M. FitzGerald and J. Knight (eds.): *High Resolution Morphodynamics and Sedimentary Evolution of Estuaries.* 2005　　　　　　　ISBN 1-4020-3295-1
9. M.D. Max, A.H. Johnson and W.P. Dillon: *Economic Geology of Natural Gas Hydrate.* 2006　　　　　　　　　　　　　　　　　ISBN 1-4020-3971-9
10. Nick Harvey (ed.): *Global Change and Integrated Coastal Management: The Asia-Pacific Region.* 2006　　　　　　　　　　　　　ISBN 1-4020-3627-2
11. N. Mimura (ed.): *Asian-Pacific Coasts and Their Management. States of Environment.* 2008　　　　　　　　　　　　　　　　　ISBN 978-1-4020-3626-2
12. J. Patrick Doody: *Saltmarsh Conservation, Management and Restoration.* 2008
　　　　　　　　　　　　　　　　　　　　　　ISBN 978-1-4020-4603-2